T0134557

Intelligent Systems Reference Library

Volume 187

Series Editors

Janusz Kacprzyk, Polish Academy of Sciences, Warsaw, Poland
Lakhmi C. Jain, Faculty of Engineering and Information Technology, Centre for
Artificial Intelligence, University of Technology, Sydney, NSW, Australia;
KES International, Shoreham-by-Sea, UK;
Liverpool Hope University, Liverpool, UK

The aim of this series is to publish a Reference Library, including novel advances and developments in all aspects of Intelligent Systems in an easily accessible and well structured form. The series includes reference works, handbooks, compendia, textbooks, well-structured monographs, dictionaries, and encyclopedias. It contains well integrated knowledge and current information in the field of Intelligent Systems. The series covers the theory, applications, and design methods of Intelligent Systems. Virtually all disciplines such as engineering, computer science, avionics, business, e-commerce, environment, healthcare, physics and life science are included. The list of topics spans all the areas of modern intelligent systems such as: Ambient intelligence, Computational intelligence, Social intelligence, Computational neuroscience, Artificial life, Virtual society, Cognitive systems, DNA and immunity-based systems, e-Learning and teaching, Human-centred computing and Machine ethics, Intelligent control, Intelligent data analysis, Knowledge-based paradigms, Knowledge management, Intelligent agents, Intelligent decision making, Intelligent network security, Interactive entertainment, Learning paradigms, Recommender systems, Robotics and Mechatronics including human-machine teaming, Self-organizing and adaptive systems, Soft computing including Neural systems, Fuzzy systems, Evolutionary computing and the Fusion of these paradigms, Perception and Vision, Web intelligence and Multimedia.

** Indexing: The books of this series are submitted to ISI Web of Science, SCOPUS, DBLP and Springerlink.

More information about this series at http://www.springer.com/series/8578

Dipti Kapoor Sarmah · Anand J. Kulkarni ·
Ajith Abraham

Optimization Models
in Steganography Using
Metaheuristics

Springer

Dipti Kapoor Sarmah
Symbiosis Institute of Technology
Symbiosis International
(Deemed University)
Pune, Maharashtra, India

Anand J. Kulkarni
Symbiosis Institute of Technology
Symbiosis International
(Deemed University)
Pune, Maharashtra, India

Ajith Abraham
Scientific Network for Innovation
and Research Excellence
Machine Intelligence Research Labs (MIR)
Auburn, WA, USA

ISSN 1868-4394   ISSN 1868-4408   (electronic)
Intelligent Systems Reference Library
ISBN 978-3-030-42046-8   ISBN 978-3-030-42044-4   (eBook)
https://doi.org/10.1007/978-3-030-42044-4

This Springer imprint is published by the registered company Springer Nature Switzerland AG
The registered company address is: Gewerbestrasse 11, 6330 Cham, Switzerland

Preface

In recent times, a high level of information security is attained by combining the concepts of cryptography and steganography along with nature-inspired optimization algorithms. One of the important media used for steganography is the Joint Photographic Expert Group (JPEG) image. An Artificial Intelligence (AI) based socio-inspired optimization algorithm referred to as Cohort Intelligence (CI) is extensively explored in the steganography approach discussed in this book along with Cognitive computing (CC) and a Multi Random Start Local Search (MRSLS) optimization algorithm. Considering four important aspects of steganography techniques—picture quality, high data hiding capacity, secret text security and computational time, extensive research efforts have been presented through four novel image-based steganography techniques/approaches based on JPEG compression.

The first approach combines CI with CC (CICC) and applied to the 8×8 quantization table. In the second approach modified MRSLS (M-MRSLS) optimization algorithm was developed and applied to the 8×8 quantization table. The third approach was proposed using the CICC and M-MRSLS optimization algorithm and applied to the 16×16 quantization table. Furthermore, CI was modified to address the practical aspects of the real world, referred to as Improved CI in the fourth approach. Improved CI has been applied to 8×8 as well as a 16×16 quantization table. The CICC, M-MRSLS and Improved CI were developed as cryptography techniques and employed to generate the optimized ciphertext. Six greyscale images of size 256×256 with different variety and texture were considered for testing the proposed algorithms. CICC and M-MRSLS optimization algorithms employed for JPEG image steganography with an 8×8 modified quantization table have demonstrated satisfactory performance against image quality and secret text capacity. The objective of using a modified 16×16 quantization table for CICC and M-MRSLS optimization algorithms was to enhance the embedding capacity of secret text compared to the 8×8 quantization table and also retained the satisfactory image quality, which was reasonable in 8×8 quantization table. In order to reduce the computational cost, Improved CI was applied to greyscale images of size 256×256. The security analysis of the proposed methods was also accomplished by applying common steganalysis attacks such as a visual attack, structural attack, and statistical attack. For experimentation, the methods CI, CICC and Improved CI used with a 16 × 16 modified quantization table were selected. The analysis of the results underscores the attainment of adequate security.

The mathematical level in all the chapters is well within the grasp of the scientists as well as the undergraduate and graduate students from the engineering and computer science domains. The reader is encouraged to have a basic knowledge of mathematical analysis. Every technique presented in the book has been coded in MATLAB software. All the executable codes could be made available on request.

This book is an outgrowth of the extensive work of over 4 years by the authors. Over this period, the algorithms have been tested extensively and the performance has been validated by solving numerous examples/problems. These illustrative examples may allow the reader to gain further insight into the associated concepts. The methods and results have been published in various prestigious journals and conferences. The suggestions and criticism of various reviewers and colleagues significantly influenced the way the work has been presented in this book.

Pune, India   Dipti Kapoor Sarmah
Pune, India   Anand J. Kulkarni
Auburn, USA   Ajith Abraham

Contents

Abbreviations

ACO	Ant Colony Optimization
AES	Advanced Encryption Standard
AIP	Analog Image Processing
ANN	Artificial Neural Network
AR	Augmented Reality
ASO	Anarchic Society Optimization
BA	Bees Algorithm
CBIR	Content-Based Image Retrieval
CBVIR	Content-Based Visual Information Retrieval
CC	Cognitive Computing
CCD	Charge-Coupled Device
CEA	Cultural Evolution Algorithm
CI	Cohort Intelligence
CICC	Cohort Intelligence with Cognitive Computing
CMOS	Complementary Metal-Oxide-Semiconductor
DA	Differential Evolution
DCT	Discrete Cosine Transform
DES	Data Encryption Standard
DFT	Discrete Fourier Transform
DIP	Digital Image Processing
DPCD	Digital Pulse Code Demodulation
DPCM	Digital Pulse Code Modulation
DSA	Digital Signature Algorithm
DWT	Discrete Wavelet Transform
EAs	Evolutionary Algorithms
EC	Entropy Coding
ECO	Election Campaign Optimization
EMD	Exploiting Modification Direction
FA	Firefly Algorithm
FB	Frame Building

FFT	Fast Fourier Transform
GA	Genetic Algorithm
GLM	Gray Level Modification
HC	Huffman Coding
HE	Honey Encryption
HMAC	Hash-based Message Authentication Code
IA	Ideology Algorithm
ICA	Imperialist Competitive algorithm
IDCT	Inverse Discrete Cosine Transform
JPEG	Joint Photographic Expert Group
JQTM	Joint Quantization Table Modification
K-MCI	K-means with Modified Cohort Intelligence
LCA	League Championship Algorithm
LDA	Linear Discriminant Analysis
LoG	Laplacian of Gaussian
LSB	Least Significant Bit
MBNS	Multiple Base Notational System
MCI	Modified Cohort Intelligence
MCIA	Multi-Cohort Intelligence Algorithm
MD	Message Digest
M-MRSLS	Modified Multi Random Start Local Search
MPP	Minimum Perimeter Polygons
MRSLS	Multi Random Start Local Search
MSE	Mean Square Error
OLSBS	Optimal Least Significant Bit Substitution
PC	Probability Collectives
PCA	Principal Component Analysis
PPM	Pixel-Pair Matching
PSNR	Peak Signal to Noise Ratio
PSO	Particle Swarm Optimization
PVD	Pixel Value Differencing
PVP	Pixel Value Prediction
QBE	Query by Example
QBIC	Query by Image Content
QKD	Quantum Key Distribution
RF	Relevance Feedback
RLD	Run Length Decoding
RLE	Run Length Encoding
RSA	Rivest Shamir Adlemen
SCO	Society and Civilization Optimization algorithm
SELO	Socio Evolution & Learning Optimization
SEO	Social Emotional Optimization
SGO	Social Group Optimization
SHA	Secure Hash Algorithm
SI	Swarm Intelligence

SIFT	Scale-Invariant Feature Transform
SIO	Socio Inspired Optimization
SLC	Soccer League Competition
SLO	Social Learning Optimization
SR	Semantic Retrieval
SURF	Speeded up Robust Features
TLBO	Teaching-Learning based Optimization

Chapter 1
Introduction

In recent times, digital communication plays a critical role in the day to day life. With the massive growth of the digital world, there is always a concern about information security. In order to understand the security aspects, one needs to recognize the importance of signal processing [12]. This is an area which deals with various operations of signals such as storing, transmitting, and filtering, etc. A signal is a well-known term used in electronics and telecommunications to carry information for an event. It can be represented as a kind of variation or any observed change in the occurrence, for example: a sound produced while speaking is referred to as speech signals, the heart functioning can be recognized as waves represented by electro-cardiograms, television and mobile generates a signal in communication field, the two dimensional signals dealing with image referred to an image signal, etc. The functioning of an image signal deals only with the various types of images and its processing; where an image is required as input and output. This field is referred to as Image Processing and is divided into two parts: (i) Analog Image Processing (AIP) and (ii) Digital Image Processing (DIP). AIP works on two-dimensional analog or electrical wave signals, for example, photographs, television images (broadcasting through dish antenna), paintings, etc. It is considered a costly and time-consuming technique as it is continuous in nature. Also, a signal distortion while processing takes lots of time to rectify. On the other side, DIP deals with digital images that work on a matrix of pixels (the combination of discrete values of 0's and 1's). It is quite easy to manipulate digital images as there is a range of algorithms [113] available for processing. DIP is considered one of the active research areas [124] and helped researchers to solve different types of practical problems [45, 107, 109] where visual observations are involved such as analyzing and manipulating the images.

The next Sect. 1.1 of this chapter explains the basics of DIP with practical examples. The objectives and various algorithms used for processing and manipulating images are elaborated in Sect. 1.2. The techniques employed for DIP are also explained in Sect. 1.2.1. Further, the relevant practical and the latest real-world applications are expressed in Sect. 1.3. The summary is mentioned in Sect. 1.4.

© The Editor(s) (if applicable) and The Author(s), under exclusive license to Springer
Nature Switzerland AG 2020
D. K. Sarmah et al., *Optimization Models in Steganography Using Metaheuristics*,
Intelligent Systems Reference Library 187,
https://doi.org/10.1007/978-3-030-42044-4_1

1.1 The Basics of Digital Image Processing (DIP)

Digital Image Processing (DIP) performs various operations on images for different classes e.g. image enhancement [49], image analysis [65], image compression [75], image transformation [59], etc. The image enhancement is used to manipulate the image for extracting the required and important key features as identified by a user, for example: adjusting contrast value for an image. The meaningful information is automatically produced using different tools such as neural networks [117]; edge detectors [13, 97], etc., are referred to as Image analysis. The common examples of Image analysis are Image segmentation [65], motion detection [36], object detection [51], etc. Image compression is also considered one of the important classes of DIP for converting a source image file to a resultant image that consumes less memory space. The image compression methods are explained in Sect. 1.2.1 in detail. The next class of DIP is the image transformation in which a function is applied to an input image of one domain to produce an output image of another domain, for example, Cosine transforms [76] and Fourier transforms [9]. The different domains used for DIP are also explained in the next Sect. 1.2.1. DIP is employed in a variety of applications such as medical imaging [26], information security [85], remote sensing [34], document processing [5], textiles [7], military [79], computer vision [107], deep learning [118, 127], etc. Nowadays, its usage in practical applications has vastly increased.

There is a need to understand the concept of images in order to execute DIP techniques which are specifically applied to images. The images can be considered as two-dimensional signals of continuous variables x and y, where x and y denote the two coordinates in the horizontal and vertical axis respectively. The mathematical function $g(x, y)$ is used to calculate the image value at a particular location with respect to the coordinates. Sampling is required in order to convert the image data from continuous to a digital form. Further, the image needs to be transformed into a two-dimensional matrix having some numbers which are quantized further to form integer values from fine accurate values. To apply various operations on images, DIP manipulates those fine accurate values which help to identify the color or gray level of the image. The number of pixel values of the image identifies its resolution referred to a term spatial resolution. The images having different spatial resolutions are depicted in Fig. 1.1. The visibility of a low-resolution image of 640×480 pixels, for instance, could be excellent while displaying on the internet. However, it might appear distorted

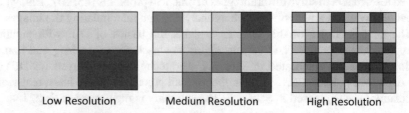

Low Resolution   Medium Resolution   High Resolution

Fig. 1.1 Spatial resolution (pixel size)

when printed or enlarged. High-Resolution images such as 1280×1024 pixels look good most of the time in internet and print form as it contains sufficient pictorial information such as fine picture details, color information, etc.

There are $2^8 = 256$ gray levels considered for an 8 bit image. The brightness value of a grayscale image referred to as pixel intensity, changes from 0 to 255, where 0 represents black and 255 represents white pixel value. Whereas, for a color image, there are 3 values between 0 and 255 per pixel represented as RGB.

In the next section, the objectives and various techniques used for DIP is identified and elaborated.

1.2 Objectives and Methods Used for Digital Image Processing (DIP)

There is a crucial role of computer algorithms in DIP as they are employed to solve various tasks such as digital image detection, image compression, image enhancement, image estimation, etc., as mentioned in Sect. 1.1. The objective of using DIP is divided into five categories: (i) Visualization process [68], (ii) Image sharpening and restoration [70], (iii) Image retrieval [30], (iv) Pattern measurement [127] and (v) Image recognition [104] as depicted in Fig. 1.2. The first category of visualization process perceives and creates the images or visuals for effective communication. This has been widely used from the ancient times in a form of cave paintings, technical drawings, data plots, etc., to computer graphics, animations, etc. DIP improves the visualization process in various applications i.e. scientific visualization [106], educational visualization [35], information visualization [16], visual communication [100], visual analytics [20], etc.

The second category, Image sharpening, and restoration help to create a better image by focussing on identified features, adjusting the contrast between bright and dark regions, reducing noise, upgrading camera focus, decreasing motion blur, etc. Image sharpening is explicitly used to improve image description such as borders, corners, contrast, edge, intensity, etc. As the name suggests, the image restoration is employed to re-establish the noisy image to its clean and original form by removing

Fig. 1.2 Objectives of digital image processing (DIP)

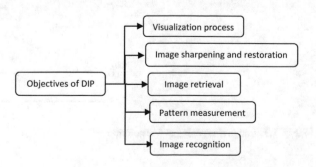

scratches, blur, etc. A few of the identified features of an image under this category
are depicted in Fig. 1.3. The grayscale image of 'baby' of size 256×256 is considered
to showcase a few of the features of the second category, which is implemented in
Matlab R2017. One can easily perceive the difference between different features as
shown in the figure.

Image retrieval, the third category utilizes various methods for browsing, seeking
and retrieving images from an image database [22], for example, Cvonline: Image

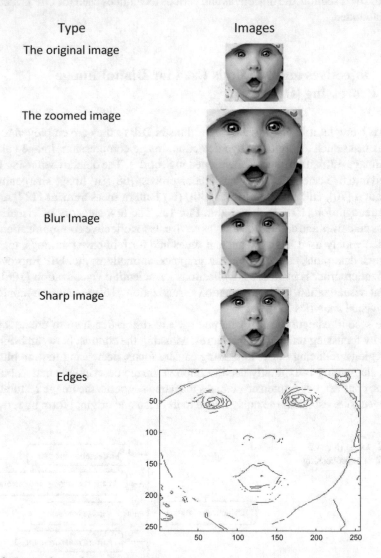

Fig. 1.3 Image sharpening and restoration

Databases, ImageProcessingPlace, etc., as mentioned in Refs. [33, 34]. The traditional methods make use of annotation words determined by adding metadata to the images. The metadata could be in the form of caption, keyword, descriptions, etc. In order to save time and effort, several web-based image annotation tools are developed for automatic image annotation. A few popular open-source tools are LabelMe [93], VGG Image Annotator [6], Sloth [93], Labelbox [60], Fast Annotation Tool [93], etc. The criteria for selecting such tools are based on its price, functionalities, and format. The other method of image retrieval is based on content or low-level features similarity [57] instead of textual information such as color, shape, texture, etc. This method is referred to as Content-based Image Retrieval (CBIR) [78] which could be also familiar as Query by Image Content (QBIC) [33] and Content-based Visual Information Retrieval (CBVIR) [77]. CBIR is mostly used for computer vision applications in Crime prevention [53], Intellectual property [19], Journalism and advertising [32], Education and training [27], Web searching [114], etc. However, there are certain challenges associated with the CBIR method and one of the major challenges is the semantic gap problem [102] between low-level content and high-level content. To understand the requirement and improve retrieval research, the CBIR system employed different techniques based on the type of user queries, for example, Query by Example (QBE) [83], Semantic Retrieval (SR) [91], Relevance Feedback (RF) [2], etc. The well-known CBIR image retrieval systems are Query by Image Content (QBIC) [89], Visual Seek [103], Blobworld [15], MARS [81], etc. The researchers have also shown their great interest in a better image retrieval system by considering the enormous use of digital images in a variety of business applications and also proposed several techniques based on neural networks [18].

The objective of image processing as mentioned in the fourth category i.e. pattern measurement is to identify the repeatable objects which appear regularly in the image data based on certain key features. In order to get more clarity, the pictorial view of a binary image is depicted in Fig. 1.4. The binary image consists of only two values 0 and 1 for each pixel for measuring the pattern.

The last category i.e. image recognition helps to identify certain features or objects in a digital image. This term can be correlated with computer vision as this concept mainly employs various DIP algorithms to recognize the objects in a variety of images. Recently, the concepts of image recognition and computer vision became very popular amongst researchers [54, 115] due to the advancements of machine

Fig. 1.4 Pattern measurement

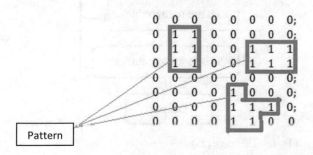

learning and deep learning concepts. Also, the combination of these concepts helped to solve a variety of real-time problems such as face recognition [50], categorizing objects [31], identifying pedestrians [67], security surveillance [90], vehicles on the road [88], etc. The well-known examples of image recognition systems are Tensor Flow by Google [80], Deep Face by Facebook [39], Project Oxford by Microsoft [110], License plate matching [87], Optical character recognition [28], etc. There are certain challenges associated with image recognition in the real-world such as viewpoint difference [46], scale difference [1], deformation and inter-class difference [69], which are still open for researchers.

Generally, there are seven steps involved in DIP from accepting an image as an input and generating the desired image output as depicted in Fig. 1.5. These seven steps are: (a) Image acquisition [55], (b) Image formation [55], and pre-processing [63], (c) Image compression [72], (d) Image segmentation [74], (e) Image representation and description [40], (f) Image recognition [92], (g) Image interpretation

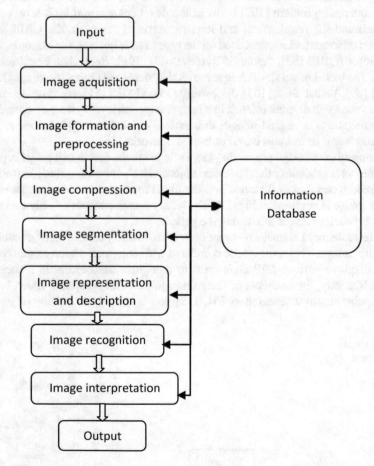

Fig. 1.5 DIP steps [25]

[52]. All the classes as discussed above and the techniques of DIP revolve around these steps. In image acquisition, an input image needs to be acquired first before applying any DIP technique. The second step prepares the input image data for the next steps by applying certain operations i.e. reducing noise, highlighting important features, geometrical transformations, etc. The main aim of image formation and pre-processing is to improve the image data before passing to any DIP algorithms. Image compression is a technique that helps to reduce the memory space required for storing an image which in turn fastens its transmission rate for transferring from sender to receiver. The capability of Image segmentation as the name suggests is used to segment the image into different objects. In image representation, the image input data is converted into a form required for further processing. It includes the color information, structure of an image, the type of an image, e.g. bitmap image, etc. In order to segment the objects in an image, image description is required which focuses on extracting the similar key features as quantitative information to distinguish the objects from one another.

Image recognition helps to locate persons, objects, animals, etc. in images. This assigns a label to each object as per the stored information. In the last step, image interpretation is required for analyzing the image output on the basis of color, size, shape, texture, shadow, pattern, etc. The required information about the problem input can be gathered from an information database as shown in Fig. 1.5.

1.2.1 Techniques Used in DIP

There are several techniques involved in order to execute each step of Fig. 1.5.

(a) The first step of image acquisition captures an image by using different cameras based on the applications. The cameras which are receptive to infrared radiation, x-rays, and visual spectrum can be used for capturing infrared images, x-ray images, and normal grayscale or color images respectively. The concept of Image acquisition depends on three steps:

(i)   An optical device to focus on energy
(ii)   Reflection/Absorption of energy by elements of objects
(iii) A sensor to measure the amount of energy

Basically, this process depends on the hardware device i.e. a camera and a sensor inside the camera. A sensor plays an important role in image acquisition as it converts light into electrical charges. Mostly, the digital cameras use image sensors referred to as charge-coupled device (CCD). Sometimes, the technology of complementary metal-oxide-semiconductor (CMOS) can also be used as an image sensor. CCD devices are mature enough to produce high quality and low noise images. However, they consume lots of power in comparison to CMOS devices.

(b) Once the image data is acquired, the second step of image pre-processing is used to prepare images for further analysis according to requirements of applications such as character recognition, texturing an image, etc. The intelligent image pre-processing/enhancement techniques employ a certain process to apply different filter operations on Image referred to as convolution. Also, it helps to construct the problem outcome in an effective way. The image pre-processing techniques could be divided into the following categories as depicted in Fig. 1.6.

(i)   Spatial domain image filtering operations,
(ii) Frequency domain image filtering operations, and
(iii) Morphological processing.

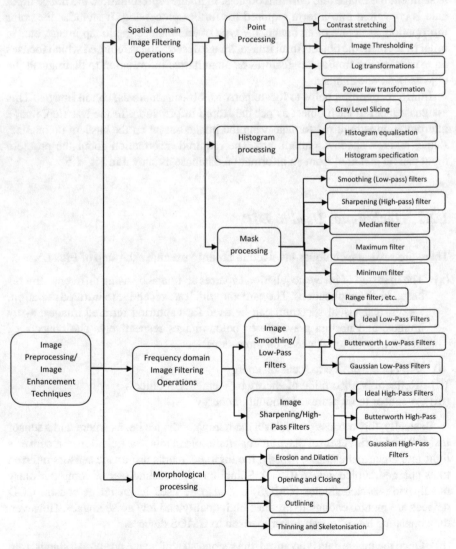

Fig. 1.6 Image pre-processing techniques

The spatial domain image filtering [63] operations are used to modify the intensity value of the current pixel by observing its neighborhood pixel values which could be further partitioned in Point processing [63], Histogram processing [17], and Mask processing [63]. Point processing is considered one of the simplest transformations in the spatial domain and categorized in methods such as Contrast Stretching [82], Image Thresholding [63], Log Transformations [3], Power Law Transformation [3], and Gray Level Slicing [84].

The contrast stretching referred to as normalization used to stretch the range of intensity values to improve the contrast of an image, for example, digital x-ray images in medical imaging as shown in Fig. 1.7, which shows the difference between low contrast and high contrast.

The image thresholding is used to segment the objects into two classes i.e. foreground and background. This is one of the important techniques employed for image segmentation as depicted in Fig. 1.9. For example, Global image thresholding [63] is done by Otsu's method [119] as shown in Fig. 1.8. The grayscale image 'baby'

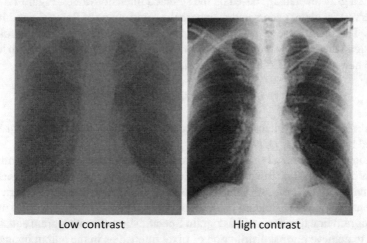

Low contrast   High contrast

Fig. 1.7 Chest X-ray image stretching

Fig. 1.8 Global thresholding

having size 256×256 is considered for image thresholding in Matlab R2017.

The next category of log transformations is used to replace each pixel value of an image by its logarithmic values to enhance the contrast of lower intensity values. It helps to narrow the brighter range of pixel values and expands the dark pixels. One can apply this transformation when there is a need to reduce the skewness distribution of image for better interpretation as shown in Fig. 1.9. On the other hand, power-law transformations are employed to highlight the objects from lighter to a darker image and can be used by the expression: $s = c \times r^{\gamma}$ where s and r are the pixel values of the output and input image respectively, c is a constant value and γ is referred to as gamma value. In order to reduce the computer monitor display problems for varying intensity values, different gamma values are used in this transformation as showcased in Fig. 1.10 for better understanding.

The gray level slicing also referred to as Intensity slicing and is one of the important techniques for emphasizing a particular range of gray levels or intensity values of an image by diminishing or by leaving the rest of the values. It could be used to segment a gray level image based on the selected intensity values. Figure 1.11 could be observed for better understanding where the same image of Baby of size 256×256 is used. The intensity level selected for gray level slicing is based on the following expression where $b(i, j)$ is referred to as image intensity values at ith and jth location and the new value i.e. 255 is copied to $c(i, j)$.

$$if \ b(i, j) \geq 150 \ then \ c(i, j) = 255 \tag{1.1}$$

Here, all the image intensity values more than 150 are considered as a group and replaced by 255 in $c(i, j)$.

Histogram processing is another method under spatial domain filtering and is used to visualize the tonal distribution of a digital image in a graphical way. It enables to distribute the number of pixel values of the image in a graph based on its corresponding intensity values. Histogram processing could be done by applying the methods of Histogram Equalisation and Histogram Specification. The histogram equalization is used to achieve normal distribution of pixel intensities in the output image which

Original Image   **Log Transformation Image**

Fig. 1.9 Logarithmic transformation

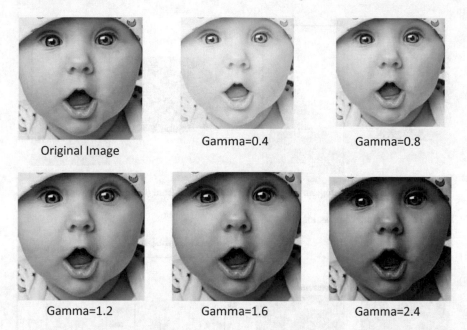

Fig. 1.10 Power law transformations

Fig. 1.11 Gray level slicing

enables to improve the contrast of the image. This method adjusts the intensity values in the image histogram as shown in Fig. 1.12. On the other side, the histogram specification method referred to as histogram matching is employed to remap the pixel intensity values to match the existing histogram with a specified histogram of a reference image to achieve the output image. The graphical representation of this method is shown in Fig. 1.13 for more clarity.

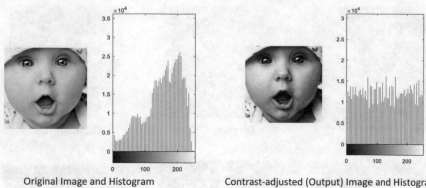

Original Image and Histogram   Contrast-adjusted (Output) Image and Histogram

Fig. 1.12 Histogram equalisation

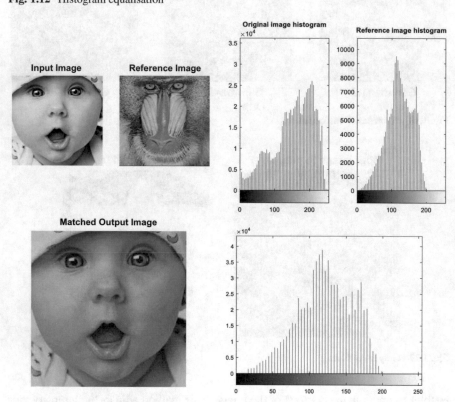

Fig. 1.13 Histogram matching

This method is implemented in Matlab by considering two images: RGB image of Baby, size 512×512 and a reference RGB image of Baboon, size 256×256. The mapped output RGB image of Baby, size 512×512 and its histogram is depicted in Fig. 1.13.

Mask processing is also one of the techniques under spatial domain image filtering in which an image mask is applied to cover certain features of the image. This process could be done by setting some of the pixel values of the image to zero. There are various methods of applying mask processing to the image. A few common methods are Smoothing (Low-pass) filters [23], Sharpening (High-pass) filter [23], Median filter [125], Maximum filter [42], Minimum filter [42], Range filter [113], etc. To understand these methods in a clearer way, the pictorial view is presented in Fig. 1.14 so that one could analyze the difference between these filters. The smoothing or low-pass filter is used to reduce the sharpness in intensities. It allows low frequencies or a range of frequencies and suppresses high frequencies. Such filters can be used by taking the mean or median values of the pixels. The common examples which use low-pass filters are audio amplifiers, equalizers, blurring of digital images, etc. The high-pass filter is used to sharpen the image. It suppresses the low frequencies and allows passing the high frequencies to provide finer details of the image. This is generally used in loudspeakers to reduce the low-level noise, equalizers, audio amplifiers, image modifications, noise reduction, etc.

The second category of Image pre-processing techniques makes use of image filtering operations in the frequency domain, partitioned in two groups: Low-Pass filters and High-Pass filters employed for Image smoothing and Image sharpening, respectively. In the spatial domain, filtering could be easier to understand, however, in the transform domain, it could be applied faster particularly for large images. Based on the requirement and application, these two filters can be applied to images. The frequency-domain filtering is done using a Fast Fourier Transform (FFT) [112]. Both the groups i.e. low pass filters and high pass filters are further separated into three classes: Ideal Low-pass filters [112], Butterworth Low-pass filters [112], and Gaussian Low-Pass filters [112]. The filtering operations are applied on the 'baby' image of size 256×256 in Matlab by employing these classes along with high boost filtering using Gaussian and butter-worth high pass filters as presented in Fig. 1.15 for more clarity. The cut-off radius or threshold value is considered 10 for all the cases. The boost factor is considered 1.75 for high boost filtering.

(c) The third step in DIP is image compression which allows a digital image to be stored in less memory space which could be done by two techniques i.e. Lossless compression and Lossy compression as shown in Fig. 1.16. With the help of the compression techniques, the redundant data of the image is reduced and the image can be stored effectively in space. The Lossless compression techniques allow compressing the image by retaining the same quality as of the original image. However, such images take more space. For example RAW, PNG, BMP image types, etc.

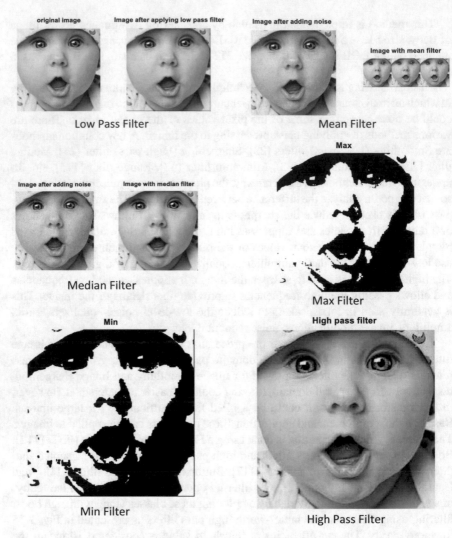

Fig. 1.14 Mask processing filters

On the other side, the Lossy compression techniques discard the redundant part of the image data and reduce the image quality to a certain extent with a high compression rate. For example JPEG, GIF image types, etc. There are various compression techniques under Lossless and Lossy as depicted in Fig. 1.16. The Lossless compression techniques are Run-length encoding [111], Entropy encoding [111], Dictionary techniques [111], Area image compression [111], Predictive coding [111], and Chain coding [111]. Run-length encoding accepts strings and reduces the redundancy of the string into one unit. For example: If there is an image of red and green stripes, and there are 15 red pixels and 12 green pixels. It should be written

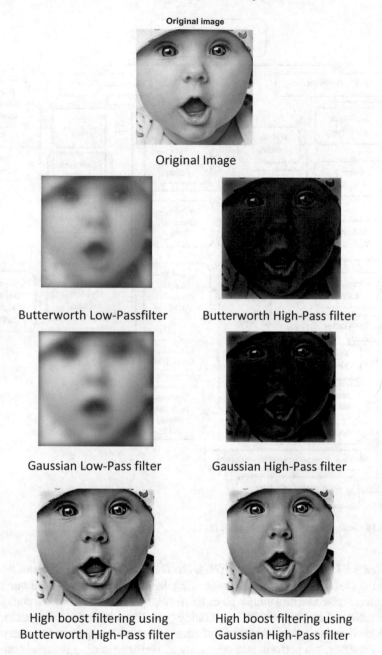

Original Image

Butterworth Low-Passfilter   Butterworth High-Pass filter

Gaussian Low-Pass filter   Gaussian High-Pass filter

High boost filtering using Butterworth High-Pass filter   High boost filtering using Gaussian High-Pass filter

Fig. 1.15 Frequency domain filters

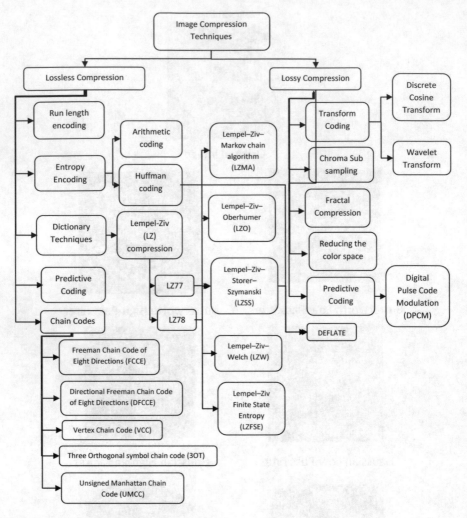

Fig. 1.16 Image compression techniques [111]

as $RRRRRRRRRRRRRRRGGGGGGGGGGGGG$, however, it could be represented as $15R12G$ using run-length encoding. In entropy coding, each unique symbol of input is replaced with a unique code. It is further partitioned into two techniques i.e. Arithmetic coding [95] and Huffman coding [95]. The arithmetic coding technique is considered better in comparison to Huffman coding with respect to the compression ratio. However, the performance efficiency of Huffman coding is found better than the arithmetic coding because of its usage of a static table for the entire process. One can apply Dictionary techniques referred to as Lempel-Ziv (LZ) compression techniques [11] for image compression. Such techniques are implemented through LZ77 [11] and LZ78 [11] algorithms. The basic idea of using dictionary-based techniques to encode the frequent symbol patterns in an effective way. The concept of static as well dynamic dictionary can be used to store a list of input strings of symbols. For example: If we consider a random house dictionary written in the English language, then the following statement: "This is an example of Dictionary-based compression", can be coded as "1/1 2/3 2/5 45/9 550/32 38/12 89/21 173/46". Each word is coded in the format of x/y where x represents the page number and y represents the word number on the page. The combination of LZ77 and LZ78 algorithms are used to propose Lempel–Ziv–Storer–Szymanski (LZSS) [66], Lempel–Ziv–Welch (LZW) [99] and Lempel–Ziv Finite State Entropy (LZFSE) [111] algorithms. Further, a new algorithm of DEFLATE [111] is proposed by combining Huffman coding and the LZSS algorithm which is generally used in portable network graphics (PNG), multiple-image network graphics (MNG) and tagged image file format (TIFF) image types. The other compression techniques are Predictive coding and Chain codes. The predictive coding can be applied for lossless as well as lossy compression and it is normally used in Digital pulse code modulation (DPCM). The chain code algorithm enables to encode each connected component or blob in the image. The well-known chaining algorithms are Freeman Chain Code of Eight Directions (FCCE) [123], Directional Freeman Chain Code of Eight Directions (DFCCE) [123], Vertex Chain Code (VCC) [122], Three Orthogonal symbol chain code (3OT) [122] and Unsigned Manhattan Chain Code (UMCC) [122]. The other type of image compression i.e. Lossy compression can be applied by using Transform coding, Chroma Subsampling, Fractal Compression, Reducing the color space and Predictive Coding (as stated earlier). One of the popular and effective techniques of Lossy compression is Transform coding [111] which can be used by employing two well-known methods such as Discrete Cosine Transform (DCT) [128] and Wavelet Transform [128]. The popular example of transform coding is the JPEG image format.

(d) The next step of DIP after image compression is applying image segmentation techniques as per the application requirement. Image segmentation is used to analyze the important parts of objects and boundaries in the image by applying to partition based on the pixel properties of the image. As depicted in Fig. 1.17, there are several image segmentation techniques that are divided into the following classes: Thresholding [63], Edge-based segmentation [58], Region-based segmentation [58], Partial differential equation-based methods [101], Clustering methods [37], Histogram-based methods [108], Trainable segmentation methods [38], and Compression-based methods [111].

The simplest method of image segmentation is considered as Thresholding which helps to convert the grayscale or color image to binary image. It could be done by separating the foreground pixels from background pixels. There are three types of commonly used thresholding techniques: Global [119], Local [119] and Multi-level thresholding [117] which could be applied to the image by using different methods. For example Otsu's method for global thresholding [119].

As shown in Fig. 1.18, the same 'baby' image as employed before for the experimentation, is segmented by Otsu's method in Global thresholding. This method considers a single threshold value for all pixels in the image. In Local thresholding,

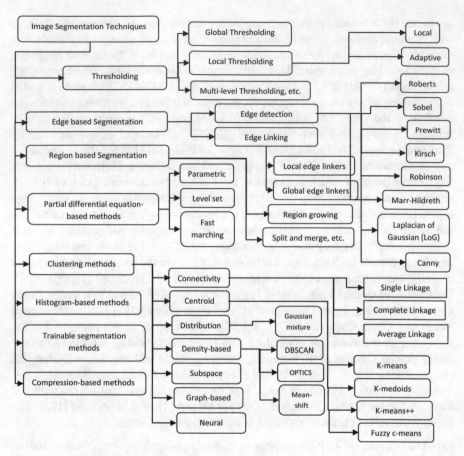

Fig. 1.17 Image segmentation techniques

Global Thresholding   Local Thresholding   Multi-level Thresholding

Fig. 1.18 Image thresholding techniques

the threshold value is changed per pixel based on its local neighborhood intensity range. The binary image is the output of Local thresholding. In Multi-level thresholding, the image is segmented in different levels based on the application. In this case, the image is partitioned into three regions by considering its level value 2. The next segmentation technique i.e. Edge-based segmentation [58] employs the number of methods to extract the information about edges in the image. Edges can lie in a boundary of two different regions where the local intensity value of a pixel is suddenly changed. It could be further categorized into Edge detection [8] and Edge linking techniques [62]. The edge detection techniques are used to identify and locate edges in the image. The well-known algorithms of this technique are Sobel, Prewitt, Kirsch, Canny, Roberts, Laplacian of Gaussian (LoG), etc. Once the pixel of an edge is identified, the edge linker accumulates those pixels together into a set of edges. This method is categorized into two classes: Local edge linkers and global edge linkers. The pictorial view of Edge-based segmentation techniques as implemented in Matlab R 2017 helps researchers to understand the methods as shown in Fig. 1.19.

The Sobel edge detector, a gradient-based method works on the principle of first-order derivatives. It uses the horizontal and vertical kernel to detect edges. The Prewitt, being the oldest method works on a similar principle of Sobel. However, their mask values are different. Robert's method is also similar to Sobel. However, it considers the mask size of 2×2 instead of 3×3. The LoG is based on one kernel and calculates second-order derivatives of the image. Though it is highly sensitive to noise, computationally faster results can be achieved. The Canny edge detection algorithm is a multistage algorithm that locates a number of edges in images.

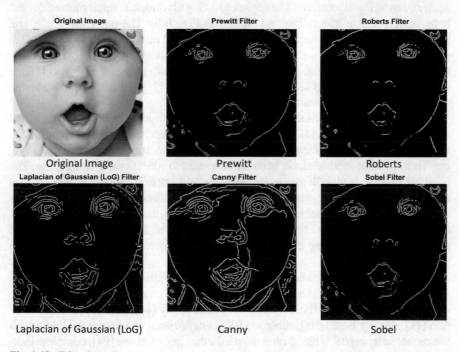

Fig. 1.19 Edge detection

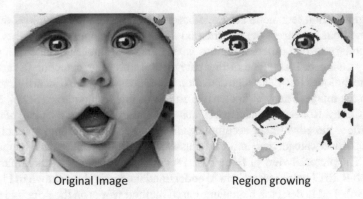

Original Image   Region growing

Fig. 1.20 Region based image segmentation

The region-based segmentation techniques are based on identifying and locating other kinds of perceptual boundaries. Such boundaries can be analyzed by considering neighborhood features and extracting those regions referred to as classes. The two key features of Region-based segmentation techniques are similarity and homogeneity. The common approaches for this technique are Region growing [58] and Split and merging algorithms. The region growing algorithm is implemented for the baby image of size 256×256 and shown in Fig. 1.20. The other important and effective technique of segmentation is based on partial differential equations (PDE) [101]. The PDE methods are already established and its concept is employed in other streams such as physics and mechanics. The well-known methods of this technique are parametric [101], level set [101] and fast marching [21]. The objects of an image can be represented as a parametric curve.

Some sampling methods are selected for contour or outline parameterizing based on Lagrangian techniques [41]. However, there are few limitations of this technique in terms of selection of sampling method, the internal geometric properties of the curve, etc. Thus, discrete methods have been incorporated to address those limitations. The level set method is an efficient parameter-free and implicit process which enables to estimate the geometric properties of the curve and allows modifying topology. In order to solve boundary value problems, fast marching methods are employed. This method is effectively used for image segmentation. The limitations of this method are reduced by improvising the previous model to the current model i.e. generalized fast marching method.

Image clustering [37] is one of the popular techniques for image segmentation. This technique has several models that utilize different methods according to the application requirement. These models are connectivity [44], centroid [24], distribution [71], density-based [98], subspace [96], graph-based and neural [96], in which the connectivity model is based on three methods: single linkage [94], complete linkage [29], and average linkage [126]. The centroid model consists of the following methods: K-means [64], K-medoids [24], K-means++ [24], Fuzzy c-means [24], etc. Further, one of the important methods under the distribution model is a Gaussian

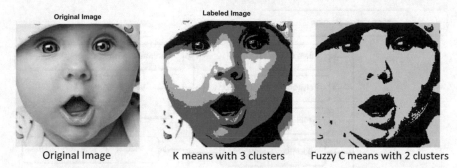

Fig. 1.21 Image clustering

mixture [71]. The DBSCAN [98], optics [37] and mean shift [37] are the important methods employed under density-based models. As there is subjectivity in performing the task of clustering, various rules are identified to understand the concept of 'similar' objects to be put in one cluster. Thus, there is a need for several models and algorithms for execution. K means and Fuzzy C means algorithm is applied to the 'baby' image in Matlab R2017 to segment to 3 and 2 clusters respectively, as shown in Fig. 1.21. The clustering techniques are not only required in image processing applications; also, they are successfully applied in data analysis, market research, pattern recognition, etc.

Further, image segmentation can also be applied using Histogram-based methods, Trainable segmentation methods and Compression-based methods. In histogram-based methods, a histogram of an image is determined. Its peaks and valleys are observed to cluster the image. The trainable segmentation methods are used to model the human knowledge for image clustering, such as neural network segmentation employing an artificial neural network (ANN) [18]. Image compression based methods are used with any of the clustering techniques where the clustering methods are applied first and then the image compression methods are used. Image clustering can also be applied by combining clustering methods in order to improve efficiency and computational time.

(e) Once the image is segmented into different regions, image representation and description techniques are utilized to combine the segmented pixels for further processing as shown in Fig. 1.22. The image representation techniques are partitioned into several classes such as Chain Codes [73], Polygonal Approximations [56], Signature [63], Boundary Segments [63], and Region Skeletons [47]. These classes can be selected based on the representation of a region: (i) boundary or external properties, (ii) texture or internal properties, (iii) both external and internal properties. The chain codes, boundary segments, polygon approximations, and signature techniques are used to represent a boundary. However, the region skeleton technique is employed to represent the structural shape of a region. There are three methods classified for polygon approximation techniques. They are minimum perimeter polygons (MPP) [56], merging [56] and splitting [56].

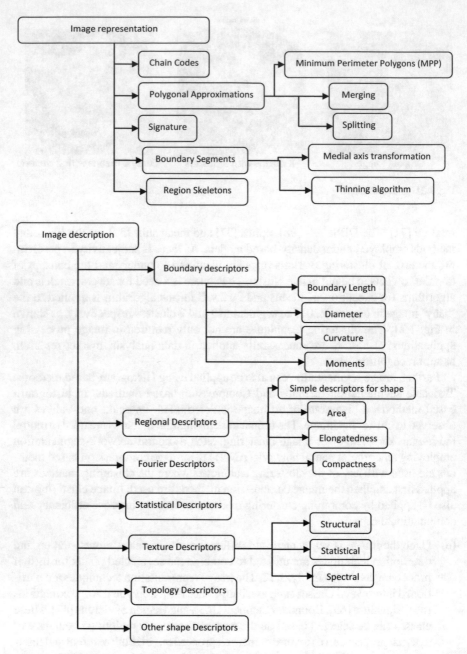

Fig. 1.22 Image representation and description techniques

The boundary segments technique is also classified into two methods: medial axis transformation [63] and thinning algorithm [63]. Further, these regions can be described depending upon their representation. For example, A boundary of a region can be described by observing its features i.e. length, width, the orientation of a line, etc. The image description techniques [92] explain the basic characteristics of an image or its objects. These descriptors are partitioned into seven categories as shown in Fig. 1.22 i.e. Boundary descriptors [43], Regional descriptors [63], Fourier descriptors [14], Statistical descriptors [61], Texture descriptors [43], Topology descriptors [10], and other shape descriptors [10]. The boundary descriptors are used to describe the boundary information of images such as boundary length, diameter, curvature and moments. Further, the regional descriptors are employed to describe information such as area, elongatedness, compactness, and shape. Furthermore, there are three categories of texture descriptors: Structural, Statistical and Spectral. The region skeleton and medial axis transformation is a part of other shape descriptors.

(f) The last step of the DIP is image recognition and image interpretation. The image recognition techniques are required to perceive objects or features in a digital image. Such techniques are useful to understand the images deeply, for example, optical character recognition, face recognition, gradient matching, etc. are some of the applications of image recognition. These techniques can also be employed for drones, manufacturing, military surveillance, autonomous vehicles, forest activities, gaming industry, etc.

To recognize the objects or its features of an image, some of the important feature extraction techniques are used: Scale-invariant Feature Transform (SIFT) [4], Speeded up Robust Features (SURF) [105], Principal Component Analysis (PCA) [48], and Linear Discriminant Analysis (LDA) [121] as shown in Fig. 1.23. Also, the image interpretation method is applied to interpret the details of the image and identifying its features, for example, size, shape, shadow, texture, pattern, location, etc.

Fig. 1.23 Feature extraction techniques

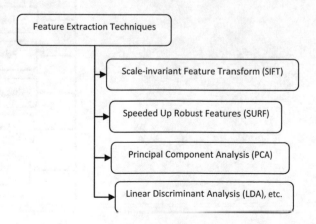

This method is generally used for aerial photographs and satellite imageries where most of the features are not located. The next section talks about a few of the current practical applications of DIP.

1.3 Current Applications

There are very wide applications of DIP and most of the technical fields are surrounded by this domain. As depicted in Fig. 1.24, some of the major real-world applications of DIP are discussed such as color processing [52], medical image analysis [45], robot vision [107], pattern recognition [127], video processing [90], remote sensing [59], feature extraction [4], image enhancement [17], digital forensics [86],

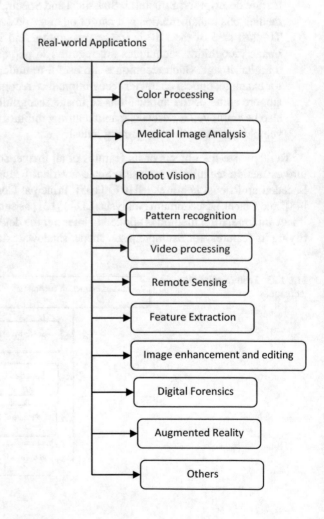

Fig. 1.24 Latest digital image processing applications

augmented reality [120] and other applications [38, 47]. Sometimes these applications employ each other's techniques to improve the computational efficiency or the outcome. For example: To click a beautiful selfie, one can use the techniques of face detection, blurring the background by maintaining a clear foreground, etc. Bar code scanning is also one of the major applications of DIP which is used for Paytm payments. To portray a panoramic view, the concept of image homography [116] is employed. Facebook, currently uses the concept of face detection techniques [50] to recognize and tag faces. Also, face recognition can be used in Mobile phones while capturing photos. Motion detection [36] can also be one of the important applications of DIP in video processing. Further, there are developments of biometric-based systems under pattern recognition, employed for students' and staff attendance [35]. The hand-writing recognition [60] is also one of the well-known and effective applications of pattern recognition in DIP. Also, there are several applications of DIP in the medical field such as X-Ray imaging, Medical CT scan, Gamma-Ray imaging, Ultraviolet (UV) imaging, etc.

Moreover, DIP is found quite effective for Robot vision in hurdle detection root, object tracking, inspecting defects for removal, defense applications, security and surveillance applications, computer-human interaction, etc. In order to analyze the satellite images, the science of remote sensing is used in which the image interpretation is done manually on image attributes such as size, shape, color, texture, shadow, pattern, etc. The digital image forensics is another area in DIP, which consists of numerous practical applications. For example photogrammetry, photographic comparison, content analysis, image authentication, etc. Further, the concept of augmented reality (AR) is becoming a vital area of image processing. The AR support images have been continuously demanded by users to have good experience in the real world. For example: (i) To experience AR in 3D models, few applications are used: AUGMENT, Sun-Seeker, etc., (ii) Argon4, AR Browser SDK, etc., web browsers can be experienced to augment the real feel, (iii) Some popular AR games such as Pokémon Go, Parallel Kingdom, Temple Treasure Hunt, Real Strike, Zombie Go, etc., are also incorporated in mobile phones, (iv) To help in spotting user location and detecting mobile device orientation, few applications of AR GPS have been considered i.e. AR GPS Compass Map 3D, AR GPS Drive/Walk Navigation, etc. It is being observed that there are various practical applications of DIP, which are not only limited to the areas mentioned in the Fig. 1.24, however, due to the advancements of the DIP concepts and by perceiving the complexities associated with the challenges of the real-world problem, this field is continuously growing in industry, academics, research, etc.

1.4   Summary

In this chapter, the authors have elaborated on the basic concepts of DIP with examples. The objectives of using DIP is also explained and justified by shedding light on the DIP methods. Further, the different techniques employed under these methods are described with examples and figures. To increase the understanding of readers (researchers, academicians, industrialists, etc.), few of the important techniques are implemented in Matlab R 2017 for the 'baby' grayscale image of size 256×256 and size 512×512. They are also showcased in different figures for visual analysis. Furthermore, the current applications of DIP are discussed. This chapter provided guidance to the readers to interpret this domain with clarity and enables them to research further in this direction.

The next chapter on Cryptography and Digital image steganography techniques discusses the basics of cryptography and image steganography in Sect. 2.1. This section is divided into two sections of 2.1.1 and 2.1.2 to elaborate on various algorithms used in cryptography and image steganography, respectively. The examples of each algorithm are also mentioned to have clarity in the concepts. The bifurcation of these algorithms is done and explained in terms of traditional and modern algorithms. Further, the concept of image compression is explained with respect to image steganography in Sect. 2.2. The various image compression models and their methods used in Lossless compression and Lossy compression are also discussed with examples under Sect. 2.2.1. One of the popular techniques of Lossy compression i.e. JPEG image compression technique is explained step by step in Sect. 2.2.2. Section 2.2.2.1 explains the state of the art for the quantization table. Further, the importance of the JPEG image compression technique is reflected in image steganography in Sect. 2.2.3. Furthermore, the latest practical applications of image steganography domain are explored in Sect. 2.3. The Chapter is summarised in Sect. 2.4.

References

1. Ahmad, A., Anisetti, M., Damiani, E., Jeon, G.: Special issue on real-time image and video processing in mobile embedded systems. J. Real-Time Image Process. **16**(1), 1–4 (2019)
2. Akuma, S., Iqbal, R., Jayne, C., Doctor, F.: Comparative analysis of relevance feedback methods based on two user studies. Comput. Hum. Behav. **60**, 138–146 (2016)
3. Al-Ameen, Z., Sulong, G., Rehman, A., Al-Dhelaan, A., Saba, T., Al-Rodhaan, M.: An innovative technique for contrast enhancement of computed tomography images using normalized gamma-corrected contrast-limited adaptive histogram equalization. EURASIP J. Adv. Signal Process. **32** (2015)
4. Azeem, A., Sharif, M., Shah, J.H., Raza, M.: Hexagonal scale invariant feature transform (H-SIFT) for facial feature extraction. J. Appl. Res. Technol. **13**(3), 402–408 (2015)
5. Baird, H.S., Tombre, K.: The Evolution of Document Image Analysis. Handbook of Document Image Processing and Recognition, pp. 63–71. Springer, Berlin (2019)
6. Barrett, D.P., Xu, R., Yu, H., Siskind, J.M.: Collecting and annotating the large continuous action dataset. Mach. Vis. Appl. **27**(7), 983–995 (2016)
7. Bennamoun, M., Bodnarova, A.: Digital image processing techniques for automatic textile quality control. Syst. Anal. Model. Simulation—Special issue: Digital signal processing and control **43**(11), 1581–1614 (2003)
8. Bhardwaj, S., Mittal, A.: A survey on various edge detector techniques. Procedia Technol. **4**, 220–226 (2012)
9. Bhattacharyya, D., Kim, T.: Image data hiding technique using discrete Fourier transformation. In: International Conference on Ubiquitous Computing and Multimedia Applications,

Ubiquitous Computing and Multimedia Applications (UCMA), Communications in Computer and Information Science book series (CCIS), vol. 151, pp. 315–323 (2011)

10. Biasotti, S., Floriani, L.D., Falcidieno, B., Frosini, P., Giorgi, D., Landi, C., Papaleo, L., Spagnuolo, M.: Describing shapes by geometrical-topological properties of real functions. ACM Comput. Surv. (CSUR) 40(4), 12 (2008)

11. Brittain, N.J., El-Sakka, M.R.: Grayscale two-dimensional Lempel-Ziv encoding. In: International Conference Image Analysis and Recognition, Lecture Notes in Computer Science (LNCS), vol. 3656, pp. 328–334 (2005)

12. Bucy, R.S.: Signal Processing. Lectures on Discrete Time Filtering, Signal Processing and Digital Filtering, pp. 55–70. Springer, New York, NY (1994)

13. Burger, W., Burge, M.J.: Edge detection in color images. In: Digital Image Processing. Texts in Computer Science(TCS), pp. 391–411 (2016)

14. Burger, W., Burge, M.J.: Fourier shape descriptors. In: Digital Image Processing. Texts in Computer Science (TCS), pp. 665–711 (2016)

15. Carson, C., Thomas, M., Belongie, S., Hellerstein, J.M., Malik, J.: Blobworld: a system for region-based image indexing and retrieval. In: Proceedings of the Third International Conference on Visual Information Systems, pp. 509–516 (1999)

16. Chaomei, C., Yue, Y.: Empirical studies of information visualization: a meta-analysis. Int. J. Hum Comput Stud. 53(5), 851–866 (2000)

17. Chen, X., Jin, J., Fei, B.: Histogram processing-based image enhancement of digital radiography for detection of cardiac calcification. In: World Congress on Medical Physics and Biomedical Engineering, May 26–31, 2012, Beijing, China, IFMBE Proceedings, 39, pp. 939–942 (2012)

18. Cristea, P.D.: Application of Neural Networks in Image Processing and Visualization. GeoSpatial Visual Analytics, NATO Science for Peace and Security Series C: Environmental Security, pp. 59–71 (2009)

19. Csurka, G.: Document image classification, with a specific view on applications of patent images. In: Current Challenges in Patent Information Retrieval, The Information Retrieval Series, vol. 37, pp. 325–350 (2017)

20. Cui, W.: Visual Analytics: A Comprehensive Overview. IEEE Access 7, 81555–81573 (2019)

21. Cui, Z., Zhang, H., Zhang, D., Li, N., Zuo, W.: Fast marching over the 2D Gabor magnitude domain for tongue body segmentation. EURASIP J. Adv. Signal Process. 190 (2013)

22. CVonline: Image Databases (http://homepages.inf.ed.ac.uk/rbf/CVonline/Imagedbase.htm)

23. Dash, S., Jena, U.R.: Multi-resolution Laws' Masks based texture classification. J. Appl. Res. Technol. 15(6), 571–582 (2017)

24. Debelee, T.G., Schwenker, F., Rahimeto, S., Yohannes, D.: Evaluation of modified adaptive k-means segmentation algorithm. In: Computational Visual Media, pp. 1–15 (2019)

25. Delp, E.J., Buda, A.J.: Digital Image Processing. Digital Cardiac Imaging. Martinus Nijhoff Publishers, pp. 5–23 (1985)

26. Deserno, T.M.: Fundamentals of medical image processing. Handbook of Medical Technology, Springer Handbooks, pp. 1139–1165 (2011)

27. Dewan, M.A.A., Murshed, M., Lin, F.: Engagement detection in online learning: a review. Smart Learn. Environ. 6(1), 1–20 (2019)

28. Din, I. U., Siddiqi, I., Khalid, S., Azam, T.: Segmentation-free optical character recognition for printed Urdu text. EURASIP J. Image Video Process. 62 (2017)

29. Dolnicar, S., Grun, B., Leisch, F.: Step 5: Extracting segments. In: Market Segmentation Analysis, Management for Professionals (MANAGPROF), pp. 75–181 (2018)

30. Duan, G., Yang, J., Yang, Y.: Content-based image retrieval research. Phys. Procedia 22, 471–477 (2011)

31. Duvdevani-Bar, S., Edelman, S.: Visual recognition and categorization on the basis of similarities to multiple class prototypes. Int. J. Comput. Vision 33(3), 201–228 (1999)

32. Faggella, D.. Machine vision for advertising—Possibilities in social and online media. EMERJ (2019). https://emerj.com/ai-podcast-interviews/machine-vision-for-advertising-possibilities-social-online-media/

33. Faloutsos, C., Barber, R., Flickner, M., Hafner, J., Niblack, W., Petkovic, D., Equitz, W.: Efficient and effective querying by image content. J. Intell. Inf. Syst. **3**(3–4), 231–262 (1994)

34. Fonseca, L.M.G., Namikawa, L.M., Castejon, E.F.: Digital image processing in remote sensing. In: 2009 Tutorials of the XXII Brazilian Symposium on Computer Graphics and Image Processing, 11–14 Oct 2009, IEEE, Rio de Janeiro, Brazil (2009)

35. Fırat, E.E., Laramee, R.S.: Towards a survey of interactive visualization for education. In: EG UK Computer Graphics & Visual Computing, Eurographics Proceedings (2018)

36. Gaba, N., Barak, N., Aggarwal, S.: Motion detection, tracking, and classification for automated video surveillance. IEEE 1st International Conference on Power Electronics, Intelligent Control, and Energy Systems (ICPEICES), 4–6 July 2016, Delhi, India (2016)

37. Galeana, D., Pacheco, H., Magadán, A.: Analysis of clustering algorithms for image segmentation and numerical databases. In: Electronics, Robotics, and Automotive Mechanics Conference (CERMA '08), 30 Sept–3 Oct 2008, IEEE, Morelos, Mexico (2008)

38. Guo, Y., Liu, Y., Georgiou, T., Lew, M.S.: A review of semantic segmentation using deep neural networks. Int. J. Multimed. Inf. Retr. **7**(2), 87–93 (2018)

39. Guo, K., Wu, S., Xu, Y.: Face recognition using both visible light image and near-infrared image and a deep network. CAAI Trans. Intell. Technol. **2**(1), 39–47 (2017)

40. Gurevich, I.B., Yashina, V.V.: Descriptive approach to image analysis: image models. Pattern Recognit. Image Anal. **18**(4), 518–541 (2008)

41. Hahn, J., Wu, C., Tai, X.: Augmented Lagrangian method for generalized TV-stokes model. J. Sci. Comput. **50**(2), 235–264 (2012)

42. Han, W., Lin, J.: Minimum-maximum exclusive mean (MMEM) filter to remove impulse noise from highly corrupted images. Electron. Lett. **33**(2), 124–125 (1997)

43. Hassaballah, M., Abdelmgeid, A.A., Alshazly, H.A.: Image features detection, description and matching. In: Image Feature Detectors and Descriptors. Studies in Computational Intelligence(SCI), vol. 630, pp. 11–45 (2016)

44. Hassanzadeh, A., Kauranne, T., Kaarna, A.: A multi-manifold clustering algorithm for hyperspectral remote sensing imagery. In: IEEE International Geoscience and Remote Sensing Symposium (IGARSS), 10–15 July 2016, IEEE, Beijing, China (2016)

45. Hatt, M., Parmar, C., Jinyi, Q., Issam, E.N.: Machine (Deep) learning methods for image processing and radiomics. IEEE Trans. Radiat. Plasma Med. Sci. **3**(2), 104–108 (2019)

46. Hayward, W.G., Zhou, G., Gauthier, I., Harris, I.M.: Dissociating viewpoint costs in mental rotation and object recognition. Psychon. Bull. Rev. **13**(5), 820–825 (2006)

47. Hu, X., Sun, B., Zhao, H., Xie, B., Wu, H.: Image skeletonization based on curve skeleton extraction. In: International Conference on Human-Computer Interaction, Human-Computer Interaction, Design, and Development Approaches. Lecture Notes in Computer Science (LNCS), vol. 6761, pp. 580–587 (2011)

48. Huan, G., Li, Y., Song, Z.: A novel robust principal component analysis method for image and video processing. Appl. Math. **61**(2), 197–214 (2016)

49. Huang, T.S.: Image enhancement: a review. Opto-electronics **1**(1), 49–59 (1969)

50. Huang, H., Hu, G.: A face detection based on face features. Fuzzy Inf. Eng. **2**, 173–180 (2009)

51. Huang, K., Wang, L., Tan, T., Maybank, S.: A real-time object detecting and tracking system for outdoor night surveillance. Pattern Recogn. **41**(1), 432–444 (2008)

52. Huang, H., Chen, Y., Hsu, W.: Integrating color, texture, and spatial features for image interpretation. In: Pacific-Rim Conference on Multimedia, Advances in Multimedia Information Processing—PCM 2004. Lecture Notes in Computer Science (LNCS), vol. 3331, pp. 327–334 (2004)

53. Idrees, H., Shah, M., Surette, R.: Enhancing camera surveillance using computer vision: a research note. Polic. Int. J. Police Strateg. Manag. **41**(2), 292–307 (2018)

54. Islam, M.T., Siddique, B.M.N.K., Rahman, S., Jabid, T.: Image recognition with deep learning. In: International Conference on Intelligent Informatics and Biomedical Sciences (ICIIBMS), 21–24 Oct 2018, IEEE, Bangkok, Thailand (2018)

55. Jähne, B.: Image Formation and Digitization. Digital Image Processing, pp. 19–52. Springer, Berlin (1995)

56. Jiang, M., Qi, X., Tejada, P.J.: A computational-geometry approach to digital image contour extraction. In: Transactions on Computational Science XIII. Lecture Notes in Computer Science (LNCS), vol. 6750, pp. 13–43 (2011)
57. Julien, C.: Automatic handling of digital image repositories: a brief survey. In: International Symposium on Methodologies for Intelligent Systems (ISMIS), Foundations of Intelligent Systems. Lecture Notes in Computer Science, vol. 4994, pp. 410–416. Springer, Berlin (2008)
58. Kaganami, H.G., Beiji, Z.: Region-based segmentation versus edge detection. In: Fifth International Conference on Intelligent Information Hiding and Multimedia Signal Processing, 12–14 Sept 2009, IEEE, Kyoto, Japan (2009)
59. Kamusoko, C.: Image transformation. In: Remote Sensing Image Classification in R. Springer Geography, pp. 67–79. Springer, Berlin (2019)
60. Kanagarathinam, K., Sekar, K.: Text detection and recognition in raw image dataset of seven segments digital energy meter display. Energy Reports **5**, 842–852 (2019)
61. Khalil, M.S., Mohamad, D., Khan, M.K., Al-Nuzaili, Q.: Fingerprint verification using statistical descriptors. Digit. Signal Proc. **20**(4), 1264–1273 (2010)
62. Kimm, H., Abolhassan, N., Lee, E.: Comparative evaluation of edge linking methods using Markov chain and regression applied heuristic. In: Iberoamerican Congress on Pattern Recognition, Progress in Pattern Recognition, Image Analysis, Computer Vision, and Applications. Lecture Notes in Computer Science (LNCS), vol. 8827, pp. 1014–1021 (2014)
63. Krig, S.: Image pre-processing. In: Computer Vision Metrics, pp. 39–83 (2014)
64. Krishnasamy, G., Kulkarni, A.J., Paramesran, R.: A hybrid approach for data clustering based on modified cohort intelligence and K-means. Expert Syst. Appl. **41**(3), 6009–6016 (2014)
65. Kuruvilla, J., Sukumaran, D., Sankar, A., Joy, S.P.: A review on image processing and image segmentation. In: International Conference on Data Mining and Advanced Computing (SAPIENCE), 16–18 Mar 2016, IEEE (2016)
66. Kwon, B., Gong, M., Lee, S.: Novel error detection algorithm for LZSS compressed data. IEEE Access **5**, 8940–8947 (2017)
67. Li, Q., Shao, C., Zhao, Y.: A robust system for real-time pedestrian detection and tracking. J. Cent. South Univ. **21**(4), 1643–1653 (2014)
68. Liu, J., Ma, W., Liu, F., Hu, Y., Yang, J., Xu, X.: Study and application of medical image visualization technology. In: International Conference on Digital Human Modeling (ICDHM), Digital Human Modeling. Lecture Notes in Computer Science book series (LNCS), vol. 4561, pp. 668–677 (2007)
69. Liu, L., Ouyang, W., Wang, X., Fieguth, P., Chen, J., Liu, X., Pietikäinen, M.: Deep learning for generic object detection: a survey. Int. J. Comput. Vis., pp. 1–58 (2019)
70. Liu, Y., Lu, W.: A robust iterative algorithm for image restoration. EURASIP J. Image Video Process. (2017)
71. Liu, X., Liao, Z., Wang, Z., Chen, W.: Gaussian mixture models clustering using Markov random field for multispectral remote sensing images. In: International Conference on Machine Learning and Cybernetics, 13–16 Aug. 2006, IEEE, Dalian, China (2006)
72. Lu, G.: Advances in digital image compression techniques. Comput. Commun. **16**(4), 202–214 (1993)
73. Lu, G.: Chain code-based shape representation and similarity measure. In: Visual Information Systems. Lecture Notes in Computer Science (LNCS), vol. 1306, pp. 135–150 (2005)
74. Mageswari, S.U., Sridevi, M., Mala, C.: An experimental study and analysis of different image segmentation techniques. Procedia Eng. **64**, 36–45 (2013)
75. Mandal, M.K.: Digital image compression techniques. In: Multimedia Signals and Systems. International Series in Engineering and Computer Science, vol. 716, pp. 169–202. Springer, Berlin (2003)
76. Mandyam, G., Ahmed, N., Magotra, N.: Lossless image compression using the discrete cosine transform. J. Vis. Commun. Image Represent. **8**(1), 21–26 (1997)
77. Marques, O., Furht, B.: MUSE: a content-based image search and retrieval system using relevance feedback. Multimed. Tools Appl. **17**(1), 21–50 (2002)

78. Mehmood, Z., Abbas, F., Mahmood, T., Javid, M.A., Rehman, A., Nawaz, T.: Content-based image retrieval based on visual words fusion versus features fusion of local and global features. Arab. J. Sci. Eng. **43**(12), 7265–7284 (2018)
79. Musheng, Y., Yu, Z.: The research of intelligent monitoring system based on digital image processing. In: Second International Conference on Intelligent Computation Technology and Automation, 10–11 Oct. 2009, IEEE, Hunan, China (2009)
80. Nguyen, G., Dlugolinsky, S., Bobák, M., Tran, V., García, A.L., Heredia, I., Malík, P., Hluchý, L.: Machine learning and deep learning frameworks and libraries for large-scale data mining: a survey. Artif. Intell. Rev. **52**(1), 77–124 (2019)
81. Ortega-Binderberger, M., Mehrotra, S.: Relevance feedback techniques in the MARS image retrieval system. Multimed. Syst. **9**(6), 535–547 (2004)
82. Panda, S.P.: Image contrast enhancement in spatial domain using fuzzy logic based interpola-tion method. In: IEEE Students' Conference on Electrical, Electronics and Computer Science (SCEECS), 5–6 Mar 2016, Bhopal, India (2016)
83. Papadias, D., Sellis, T.: A pictorial query-by-example language. J. Vis. Lang. Comput. **6**(1), 53–72 (1995)
84. Patel, B.C., Sinha, G.R.: Gray level clustering and contrast enhancement (GLC–CE) of mammographic breast cancer images. CSI Trans. ICT **2**(4), 279–286 (2015)
85. Patsakis, C., Alexandris, N.: Multimedia information security. In: Multimedia Services in Intelligent Environments, Studies in Computational Intelligence, vol. 120, pp. 257–273. Springer, Berlin (2008)
86. Peterson, G.: Forensic analysis of digital image tampering. In: IFIP International Conference on Digital Forensics, Advances in Digital Forensics, IFIP—The International Federation for Information Processing (IFIPAICT), vol. 194, pp. 259–270 (2005)
87. Prates, R.C., Cámara-Chávez, G., Schwartz, W.R., Menotti, D.: An adaptive vehicle license plate detection at higher matching degree. In: Iberoamerican Congress on Pattern Recognition, Progress in Pattern Recognition, Image Analysis, Computer Vision, and Applications. Lecture Notes in Computer Science (LNCS), vol. 8827, pp. 454–461 (2014)
88. Płaczek, B., Staniek, M.: Model based vehicle extraction and tracking for road traffic control. In: Computer Recognition Systems 2, Advances in Soft Computing, vol. 45, pp. 844–851 (2007)
89. Ragatha, D.V., Yadav, D.: Image query based search engine using image content retrieval. In: UKSim 14th International Conference on Computer Modelling and Simulation, 28–30 March 2012. IEEE, Cambridge, UK (2012)
90. Rajpoot, Q.M., Jensen, C.D.: Security and privacy in video surveillance: requirements and challenges. In: IFIP International Information Security Conference, ICT Systems Security and Privacy Protection, vol. 428, pp. 169–184 (2014)
91. Ren, F., Bracewell, D.B.: Advanced information retrieval. Electron. Notes Theor. Comput. Sci. **225**, 303–317 (2009)
92. Rosenfeld, A.: Image processing and recognition. Adv. Comput. **18**, 1–57 (1979)
93. Russell, B.C., Torralba, A., Murphy, K.P., Freeman, W.T.: LabelMe: a database and web-based tool for image annotation. Int. J.Comput. Vis. ACM **77**(1–3), 157–173 (2008)
94. Saha, P.K., Udupa, J.K., Odhner, D.: Scale-based fuzzy connected image segmentation: theory, algorithms, and validation. Comput. Vis. Image Underst. **77**(2), 145–174 (2000)
95. Salomon, D.: Data Compression. Handbook of Massive Data Sets, Massive Computing (MACO), vol. 4, pp. 245–309 (2003)
96. Saxena, A., Prasad, M., Gupta, A., Bharill, N., Patel, O.P., Tiwari, A., Er, M.J., Ding, W., Lin, C.: A review of clustering techniques and developments. Neurocomputing **267**(C):664–681 (2017)
97. Sharma, A., Ansari, M.D., Kumar, R.: A comparative study of edge detectors in digital image processing. In: 4th International Conference on Signal Processing, Computing and Control (ISPCC), 21–23 Sept. 2017. IEEE, New York (2017)
98. Shen, J., Hao, X., Liang, Z., Liu, Y., Wang, W., Shao, L.: Real-time superpixel segmentation by DBSCAN clustering algorithm. IEEE Trans. Image Process. **25**(12), 5933–5942 (2016)

99. Shim, H.J., Ahn, J., Jeon, B.: DH-LZW: lossless data hiding in LZW compression. In: International Conference on Image Processing (ICIP), 24–27 Oct. 2004. IEEE, Singapore (2004)

100. Singh, G.: Improving visual communication. IEEE Comput. Graph. Appl. **38**(1), 8–10 (2018)

101. Slíž, J., Mikulka, J.: Advanced image segmentation methods using partial differential equations: a concise comparison. In: Progress in Electromagnetic Research Symposium (PIERS), 8–11 Aug. 2016. IEEE, Shanghai, China (2016)

102. Smith, J.R.: The real problem of bridging the "Semantic Gap". In: International Workshop on Multimedia Content Analysis and Mining, Multimedia Content Analysis and Mining (MCAM), Lecture Notes in Computer Science, vol. 4517, pp. 16–17. Springer, Berlin (2007)

103. Smith, J.R., Chang, S.F.: VisualSEEK: a fully automated content based image query system. In: ACM Multimedia, pp. 97–98 (1996)

104. Song, Y., Bai, C.: Research and analysis of image processing technologies based on DotNet framework. Phys. Procedia **25**, 2131–2137 (2012)

105. Sykora, P., Kamencay, P., Hudec, R.: Comparison of SIFT and SURF methods for use on hand gesture recognition based on depth map. In: AASRI Conference on Circuits and Signal Processing (CSP 2014), AASRI Procedia, vol. 9, pp. 19–24 (2014)

106. Thomas, A.D., Maxine, D.B.: Visualization in scientific computing. Adv. Comput. **33**, 247–307 (1991)

107. Tiwari, M., Lamba, S. S., Gupta, B.: An image processing and computer vision framework for efficient robotic sketching. Procedia Computer Science, vol. 133, pp. 284–289. Elsevier (2018)

108. Tobias, O.J., Seara, R.: Image segmentation by histogram thresholding using fuzzy sets. IEEE Trans. Image Process. **11**(12), 1457–1465 (2002)

109. Tong, J., Wu, C., Chen, D.: Research and implementation of a digital image processing education platform. In: International Conference on Electrical and Control Engineering, 16–18 Sept. 2011, Yichang, China (2011)

110. Touliou, K., Maglavera, M., Ecabert, C., Pauzie, A., Willstrand, T.: SoA and benchmarking, deliverable 1.1 [Research Report]. IFSTTAR - French Institute of Sciences and Technologies of Transport, Planning and Networks, 187 (2017)

111. Uthayakumar, J., Vengattaraman, T., Dhavachelvan, P.: A survey on data compression techniques: from the perspective of data quality, coding schemes, data type and applications. J. King Saud Univ. Comput. Inf. Sci. (2018). https://doi.org/10.1016/j.jksuci.2018.05.006(Inpress)

112. Varghese, J., Subash, S., Tairan, N., Babu, B.: Laplacian-based frequency domain filter for the restoration of digital images corrupted by periodic noise. Can. J. Electr. Comput. Eng. **39**(2), 82–91 (21 April 2016)

113. Varkonyi-Kóczy, A.R.: New advances in digital image processing. Memet. Comput. **2**(4), 283–304 (2010)

114. Wenpeng, M., Minazuki, A., Hayashi, H.: Research of intelligent search engine based on computer vision. In: IEEE/ACIS 12th International Conference on Computer and Information Science (ICIS), 16–20 June 2013. IEEE, Niigata, Japan (2013)

115. Wu, J., Peng, B., Huang, Z., Xie, J.: International Conference on Computer and Computing Technologies in Agriculture. Computer and Computing Technologies in Agriculture VI, IFIP Advances in Information and Communication Technology (IFIPAICT), vol. 392, pp. 183–188 (2012)

116. Wu, Y., Tang, C., Hor, M., Liu, C.: Automatic image interpolation using homography. EURASIP J. Adv. Signal Process., 307546 (2010)

117. Xia, X., Gao, H., Hu, H., Lan, R., Pun, C.: A multi-level thresholding image segmentation based on an improved artificial bee colony algorithm. In: 2nd EAI International Conference on Robotic Sensor Networks, EAI/Springer Innovations in Communication and Computing, pp. 11–19 (2019)

118. Xin, M., Wang, Y.: Research on image classification model based on deep convolution neural network. EURASIP J. Image Video Process. **40** (2019)

119. Yang, X., Shen, X., Long, J., Chen, H.: An improved median-based Otsu image thresholding algorithm. In: AASRI Conference on Modelling, Identification and Control, AASRI Procedia, vol. 3, pp. 468–473 (2012)

120. Youm, D., Seo, S., Kim, J.: Design and development methodologies of Kkongalmon, a location-based augmented reality game using mobile geographic information. EURASIP J. Image Video Process. **1** (2019)

121. Yuan, Y., Zhao, K., Lu, H.: Multi-label linear discriminant analysis with locality consistency. In: International Conference on Neural Information Processing, Lecture Notes in Computer Science (LNCS), vol. 8835, pp. 386–394 (2014)

122. Žalik, B., Mongus, D., Liu, Y., Lukač, N.: Unsigned Manhattan chain code. J. Vis. Commun. Image Represent. ACM, **38**(C), 186–194 (2016)

123. Žalik, B., Mongus, D., Lukač, N.: A universal chain code compression method. J. Vis. Commun. Image Represent. ACM, **29**(C), 8–15 (2015)

124. Zhang, L., Zhang, L., Zhang, L.: Application research of digital media image processing technology based on wavelet transform. EURASIP J. Image Video Process. **138** (2018)

125. Zhu, Y., Huang, C.: An improved median filtering algorithm for image noise reduction. Phys. Procedia **25**, 609–616 (2012)

126. Zhu, Q., Xiong, Q., Wang, K., Lu, W., Liu, T.: Accurate WiFi-based indoor localization by using fuzzy classifier and MLPs ensemble in complex environment. J. Frankl. Inst. (In Press) (2019)

127. Zilong, H., Jinshan, T., Wang, Z., Zhang, K., Zhang, L., Sun, Q.: Deep learning for Image-based cancer detection and diagnosis—a survey. Pattern Recognit. **83**, 134–149 (2018)

128. Zixiang, X., Kannan, R., Orchard, M.T., Ya-Qin, Z.: A comparative study of DCT and wavelet-based image coding. IEEE Trans. Circuits Syst. Video Technol. **9**, 692–695 (1999)

Chapter 2
Cryptography and Digital Image Steganography Techniques

Exchange of secure information between the sender and receiver draws the attention of several researchers due to its importance in various fields ranging from national security [73] to social profile [29]. Cryptography [17] and Steganography [14, 59] are the two major streams dealing with information security. This chapter focuses on different techniques used for cryptography and steganography. The major emphasis is given on image steganography. Various popular algorithms for cryptography and image steganography are explained in Sect. 2.1. JPEG image compression and its need are also described in detail in Sect. 2.2. Quantization plays a key role in JPEG image compression used for steganography methods. As per the state of art, there can be various quantization tables having different sizes which help to increase the embedded secret text capacity. The relevant detail of such quantization tables is also presented in Sect. 2.2.

2.1 Introduction: Cryptography and Image Steganography

In Cryptography two processes are involved referred to as encryption and decryption. Encryption is a technique that transforms the secret message into an opaque form or an unreadable form referred to as ciphertext. Steganography is a process of concealing a secret message by hiding its existence. In digital steganography, a secret message is hidden into an image, video, audio, etc. Generally in steganography, the images are used as cover media because individuals often transmit digital pictures over email and other internet communication mediums. Also, after digitalization, images contain the quantization noise which provides space to embed secret data [82]. Image-based steganography uses an original image to hide the secret text, referred to as a cover image. Once the secret text is hidden into the cover image, the image with embedded text is referred to as stego image [12, 13, 48, 77, 88]. The next Sect. 2.1.1 elaborates on the cryptography algorithms. Image steganography algorithms are explained in Sect. 2.1.2.

© The Editor(s) (if applicable) and The Author(s), under exclusive license to Springer Nature Switzerland AG 2020
D. K. Sarmah et al., *Optimization Models in Steganography Using Metaheuristics*, Intelligent Systems Reference Library 187, https://doi.org/10.1007/978-3-030-42044-4_2

2.1.1 Cryptographic Algorithms with Examples

There are two types of cryptographic algorithms: Traditional and Modern as shown in Fig. 2.1. The traditional cryptography algorithms [50] are partitioned in Substitution ciphers [50] and Transposition ciphers [65]. The substitution ciphers are further categorized into Monoalphabetic [70], Polyalphabetic [70] and Polygraphic ciphers [1]. The well-known algorithms under monoalphabetic substitution are Caesar Cipher [65], Affine [50], ROT13 [50], Simple Substitution Cipher [50], etc. The Polyalphabetic cipher includes Autokey [70], Vigenere Cipher [65], One-Time Pad [3], Enigma [25], Roto cipher [25], etc. Further, popular traditional algorithms fall under the category of Polygraphic cipher are Hill [1] and Playfair [38]. The concept behind the substitution cipher techniques is to replace the letter of plain text with any other alphabetic character to form a ciphertext. The following substitution ciphers techniques are explained with an example: (a) in Caeser cipher, each letter in the alphabet is shifted to some other letter with the help of a key. For example: if the

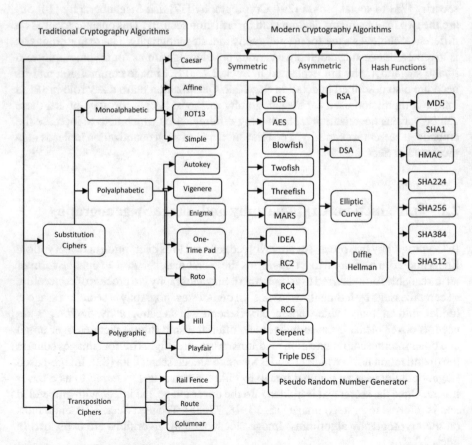

Fig. 2.1 Traditional and modern algorithms

key value is 3 and there is a letter 'a' in the plain text, then the cipher character for this letter is 'd'. This key-value can also be referred to as shift value. (b) If there are multiple shift values to be applied to several Caesar ciphers in an arrangement, then this combination forms a Vigenère cipher. (c) In a simple substitution cipher, every plaintext character is getting replaced by a different ciphertext character. For example: if the selected key for "abcdefghi" is "mqlponzux", then the ciphertext of the plain text "fade" should be "nmpo" by using this technique. The second category of traditional algorithms of Transposition ciphers consists of various popular techniques such as Rail Fence, Route, Columnar, etc. The transposition cipher technique enables us to rearrange the letters of plain text in some way. Although the traditional algorithms are easy to implement, however, they are very prone to cryptanalysis [2].

On the other side, the modern cryptography algorithms [57] are classified into three categories: (i) Symmetric [39], (ii) Asymmetric [5] and (iii) Hash functions [57]. The first category—Symmetric algorithms uses the same secret key between the sender and receiver to share the secret message. Once the secret key gets attacked by a cryptanalyst [2], a secret message can be captured by the attacker. There could be two branches for any symmetric key encryption algorithms: Stream Cipher and Block Cipher. In stream cipher [39] each character of the plain text is converted into a cipher character one by one using the secret key. On the other hand, the number of characters of a secret message makes an individual block in a block cipher. The blocks of plaintext are converted into blocks of ciphertext. The well known symmetric key encryption algorithms are as follows: Data Encryption Standard (DES) [20, 76], Advanced Encryption Standard (AES) [49, 71], Blowfish [36], Triple DES [49], Twofish [4], Threefish [4], MARS [4], IDEA [15], RC4 [4], RC6 [4], Pseudo-random number generator [4], etc. Several attacks have been identified on stream ciphers, viz.: Reused key attack [83] and Bit-flipping attack [31]. The algorithms such as DES [76], AES [49, 71], Blowfish [62], Triple DES [62], etc., fall under the branch of the block cipher [39]. Though these algorithms are complex in terms of cryptanalysis, few attacks are identified on block cipher algorithms such as Fault attacks [28], Linear and Differential attacks [18], Quantum attacks [84], etc. AES is also a well-known algorithm wherein multiple rounds are involved to transform the plain text to ciphertext. The number of rounds is selected based on the size of plain text. AES is found comparatively secure than the other mentioned algorithms due to its complex structure. However, this algorithm is also attacked by the combination of passive and active attacks [16]. The second category of Asymmetric algorithms works on two parts of keys: (a) private key, (b) public key. The well-known algorithms under this category are Rivest Shamir Adleman (RSA) [19, 80], Diffie-Hellman key exchange [89], Digital Signature Algorithm (DSA) [10], etc. The primary objective of using asymmetric key cryptography is to provide authentication and confidentiality. The third category of cryptography algorithms is hash functions which actually is the subpart of the asymmetric algorithm. It works on a restricted size of a secret message. A message digest (MD) is computed using a hash function to convert the large size of the secret text to its limited size. There could be different hash functions such as HMAC [57], MD Algorithm [63] referred to as MD4, MD5 and Secure Hash

Algorithm (SHA) [57] referred to as SHA 1, SHA 2, SHA 3, SHA224, SHA256, SHA384, SHA512, etc.

The cryptanalyst is always active in attacking cryptography algorithms to extract the secret text. There have been different cryptanalysis methods [20] such as cipher-text only attack [9], known plain text attack [30], chosen plain text attack [19], differential cryptanalysis [9], etc. On the contrary, the researchers and security specialists have also been working to make a secure cryptography algorithm to protect the secret message and to reduce the effect of cyberattacks. The outcome of the development by expert observers and researchers in this direction is Honey Encryption [85] and Quantum Key Distribution [46]. Further, various metaheuristic optimization algorithms are used as a cryptography algorithm such as Particle Swarm Optimization (PSO) [23, 40], Genetic Algorithm (GA) [27], Ant Colony Optimization (ACO) [21, 81], etc., to advance the efficiency of cryptography algorithms in terms of security.

In the next session 2.1.2, the image steganography algorithms are explained with examples.

2.1.2   Image Steganographic Algorithms with Examples

The traditional steganography algorithms [22] are based on physical, chemical or other steganography methods where the existence of the secret message is hidden and gets revealed to the intended person. Such methods are: (i) messages written on a piece of paper by secret ink, messages written on an area of postal stamps on an envelope, transmitting information via photosensitive glass during World War II, communication through sign language, etc. The modern steganography [22] uses digital messages and digital media for communication. As shown in Fig. 2.2, the Digital steganography algorithms are divided into four categories as per the type of media i.e. text, image, video, audio, and protocol. The text steganography [69] is a technique to hide a secret text message into another text file. This technique is classified into three categories: (a) Structural [26], (b) Random and statistical generation [26], and (c) Linguistic [26]. The structural steganography is also referred to as format based steganography where the cover message would not be modified. The secret text would be hidden in the spaces between the words, line, paragraphs, etc. This technique can be further divided into the following methods: (a) Open Space, (b) Line/Word shift, (c) Zero-width, (d) Feature, and (e) Emoticons. The second classification of text steganography i.e. random and statistical generation technique is employed to produce the cover text messages randomly to hide every secret message. The language structure and grammar are used to generate the proper cover message. The well-known methods of this technique are compression and random cover. Further, the third classification of Linguistic is divided into two methods. They are Syntactic and Semantic algorithms. Such algorithms use linguistic structure to hide secret text. The most commonly used digital media for steganography is image steganography where the secret text can be hidden into a cover image.

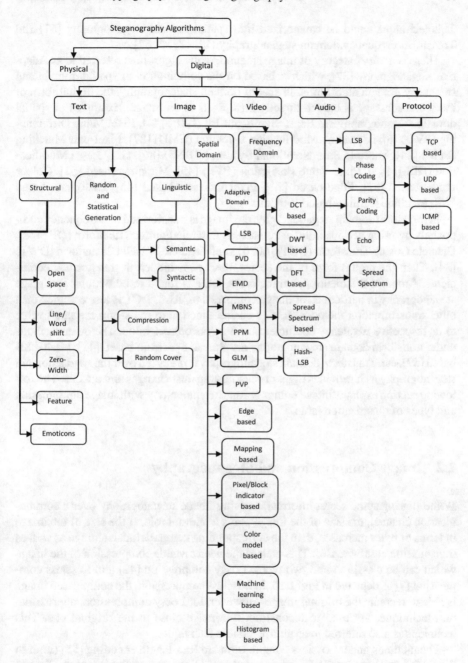

Fig. 2.2 Image steganography-traditional and modern algorithms [22]

This technique could be categorized into spatial domain steganography [67] and Transform/Frequency domain steganography [12, 13, 56, 67].

There is a third category of image steganography algorithms referred to as Adaptive steganography [37], which is based on the combination of spatial domain and frequency domain algorithms. In spatial domain steganography, the digital form of the secret message is hidden directly into pixels of an image. Examples of spatial domain steganography are Least Significant Bit (LSB) [44], Pixel Value Differencing (PVD) [6], Exploiting Modification Direction (EMD) [87], Pixel-pair Matching (PPM) [24], Multiple Base Notational System (MBNS) [86], Gray Level Modification (GLM) [45], Pixel Value Prediction (PVP) [45], Mapping based [51], Colour model-based [51], Edge-based [72], Pixel indicator based [45], Histogram based [45], Machine learning-based [90], etc.

In transform domain steganography, the image is transformed into frequency components by using various transforms such as Discrete Fourier Transform (DFT) [8], Discrete Cosine Transform (DCT) [60, 61] and Discrete Wavelet Transform (DWT) [34]. Then the digital form of the secret message is hidden in transformed coefficients. Amongst the mentioned transforms, DCT is found to be popular for image steganography in particular for image compression. Also, DCT is less complex than other transformation methods. It uses cosine function to transform the pixel value to its respective frequency coefficients and works on real value. The other methods under transform domain steganography are Spread Spectrum based [41], Hash-LSB based [57], etc. Transform domain steganography is preferred over the spatial domain steganography as it provides higher robustness against changes and attacks. The following section explains image compression, steganography with image compression and types of quantization table.

2.2   Image Compression and Steganography

While sharing some secret information using image steganography over a communication channel, the size of the image plays a relevant role. If the size of the image in terms of bytes increases, then the corresponding computational, storage as well as transmission cost increases. This creates the need towards compression of the image which can be classified into two types: Lossy compression [42] and Lossless compression [7] as depicted in Fig. 1.16. In Lossless compression, the compressed image is able to recreate the original image. However, for Lossy compression, approximation techniques are used to reconstruct the image close to the original one. This technique is also referred to as an irreversible technique.

The methods under Lossless compression are Run-length encoding [55] (refer to Sect. 1.2.1 for more detail), Entropy encoding [55], etc. The methods such as Chroma sub-sampling [43], Transform coding, etc., are few well-known methods under the lossy compression category. Lossy compression is more commonly used to compress media files whereas lossless compression is used to compress text and data files. The most popular method under Lossy compression is Transform coding which transforms the raw data into the required domain. The JPEG image compression method is an example of Transform coding and it is widely used for steganography. The detailed review of JPEG image compression with steganography is explained in the following section.

2.2.1 Steps of JPEG Image Compression

DCT based JPEG image compression [52, 35] is the most popular lossy compression technique [60] in which the redundant data is discarded and image of the size is reduced significantly. In this technique, the pixel values of the image are transformed into DCT coefficients with the help of a standard quantization table. The DCT quantized values are calculated which are used to hide the bits of the secret message. According to Bohme [11], this method is more secure as compared to the methods used under spatial domain steganography and less prone to attacks by compromising the hidden text capacity [14]. However, the major challenges of this method are to maintain the good capacity of hiding secret bits and to select quantized DCT coefficients for hiding. The details of the JPEG image compression technique are presented below and shown in Fig. 2.3.

JPEG image compression can be applied to grayscale or color image.

Compression techniques:

1. Block preparation
2. DCT
3. Quantization
4. Zig-zag scan
5. DPCM encoding
6. Run-length encoding (RLE)
7. Huffman encoding technique
8. Frame building.

An image is represented by one or more 2D array of pixels (refer to Sect. 2.1 for more detail). The block preparation step breaks each 2D array of the image into

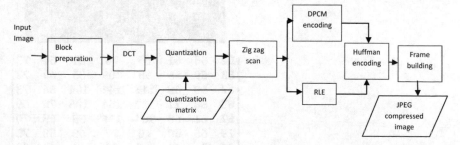

Fig. 2.3 Block diagram of JPEG compression

blocks each of size of 8 × 8 pixels per block. The block preparation and the details of the block in pixels are shown in Figs. 2.4 and 2.5, respectively.

1. DCT:

The objective of DCT [61] is to transform each block from the time domain to the frequency domain. Figure 2.6 presents a block detail of an image where DCT is applied.

2. Quantization:

The human eye cannot respond to all the coefficients. It responds primarily to the DC coefficients and lower spatial frequency coefficients [66] (Please refer to Sect. 3.2.3). This property is taken into account in the quantization phase where the higher spatial frequency coefficients are dropped in the transformed array. A division operation is

Fig. 2.4 Block preparation

Fig. 2.5 Block detail

$$
\begin{bmatrix}
52 & 55 & 61 & 66 & 70 & 61 & 64 & 73 \\
63 & 59 & 55 & 90 & 109 & 85 & 69 & 72 \\
62 & 59 & 68 & 113 & 144 & 104 & 66 & 73 \\
63 & 58 & 71 & 122 & 154 & 106 & 70 & 69 \\
67 & 61 & 68 & 104 & 126 & 88 & 68 & 70 \\
79 & 65 & 60 & 70 & 77 & 68 & 58 & 75 \\
85 & 71 & 64 & 59 & 55 & 61 & 65 & 83 \\
87 & 79 & 69 & 68 & 65 & 76 & 78 & 94
\end{bmatrix}
$$

-412	-30	-61	27	56	-20	-2	0
4	-22	-61	10	13	-7	-9	5
-47	7	77	-25	-29	10	5	-6
-49	12	34	-15	-10	6	2	2
12	-7	-13	-4	-2	2	-3	3
-8	3	2	-6	-2	1	4	2
-1	0	0	-2	-1	-3	4	-1
0	0	-1	-4	-1	0	1	2

Fig. 2.6 DCT applied to an image block

16	11	10	16	24	40	51	61
12	12	14	19	26	58	60	55
14	13	16	24	40	57	69	56
14	17	22	29	51	87	80	62
18	22	37	56	68	109	103	77
24	35	55	64	81	104	113	92
49	64	78	87	103	121	120	101
72	92	95	98	112	100	103	99

Fig. 2.7 Standard JPEG quantization matrix

performed using the defined threshold value (JPEG standard quantization matrix as shown in Fig. 2.7) as the divisor. The whole concept is self-explaining in Fig. 2.8.

3. Zig-zag scan:

The entropy coding algorithm operates on one-dimensional string or values i.e. vector. The algorithm converts a 2D array into a 1D array, the process is referred to as vectoring. A simple scan row-by-row in a definite pattern from left to right would lead to a mixture of zero or non-zero values. To cluster together zero and non-zero values a zigzag scan of the array is performed as presented in Fig. 2.9.

4. Digital pulse code modulation (DPCM) Encoding:

Digital pulse code modulation (DPCM) is a signal encoder that follows pulse-code modulation (PCM) but performs some extra functionality. The input can be either of analog signal type or it can be of digital signal type. If the input is a continuous-time analog signal, the signal has to be sampled first to get a discrete-time signal which

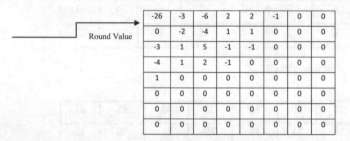

412	-30	-61	27	56	-20	-2	0
4	-22	-61	10	13	-7	-9	5
-47	7	77	-25	-29	10	5	-6
-49	12	34	-15	-10	6	2	2
12	-7	-13	-4	-2	2	-3	3
-8	3	2	-6	-2	1	4	2
-1	0	0	-2	-1	-3	4	-1
0	0	-1	-4	-1	0	1	2

$\div$

16	11	10	16	24	40	51	61
12	12	14	19	26	58	60	55
14	13	16	24	40	57	69	56
14	17	22	29	51	87	80	62
18	22	37	56	68	109	103	77
24	35	55	64	81	104	113	92
49	64	78	87	103	121	120	101
72	92	95	98	112	100	103	99

Round Value

-26	-3	-6	2	2	-1	0	0
0	-2	-4	1	1	0	0	0
-3	1	5	-1	-1	0	0	0
-4	1	2	-1	0	0	0	0
1	0	0	0	0	0	0	0
0	0	0	0	0	0	0	0
0	0	0	0	0	0	0	0
0	0	0	0	0	0	0	0

Fig. 2.8 Quantized matrix

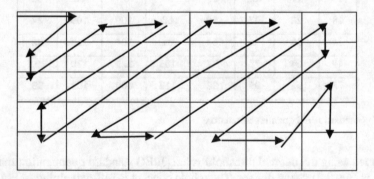

Fig. 2.9 Zig-zag scanning pattern

is then given as input to the DPCM encoder. There is one DC coefficient per block. The DC coefficient is the first and the mean value of all the 64 values of that block. The sequence of the DC coefficient is encoded in DPCM.

5. Run-length encoding (RLE):

As described in Sect. 1.2.1, Run-Length encoding (RLE) is the simplest form of lossless data compression in which streams of data are saved as a single data value and count with a special symbol to denote the meaning of the representation. This is most appropriate on data that contain the number of runs of similar sequences: for instance, simple images like icons, line drawings, fax documents, and animations. Due to the zig-zag scan, the AC coefficient of each block has been grouped together in such a way that the zero values have been clustered together. For example, if the data string is 'ttttppp'. On applying the RLE algorithm, the encoded data string will be '@t4@p3', where @ directs the decoder that the first following value is the character and its subsequent value shows the frequency of that character.

6. Huffman Encoding:

Huffman coding is an encoding algorithm that is used for lossless data compression (refer to Fig. 1.16) [68]. The term indicates the use of a variable-length code table for encryption of a source symbol (a character in an input file). The philosophy behind file reduction is, this code uses a smaller number of bits for frequently occurring characters and more number of bits for less frequent characters. Huffman coding technique is applied to both the differential encoded DC coefficients of the different blocks as well as the AC coefficient within a block.

7. Frame building:

The frame building block is the final step of the JPEG compression method which does the final assembling of the data, checks for error before sending the data to the output as an encoded data stream. The basic purpose of cryptography and steganography is to protect the secret message from unauthorized users or attackers while transmitting it through a certain communication channel. JPEG-JSTEG is a well-known commercial information hiding tool. It was proposed by D. Upham as described in [40, 74]. It replaces the least significant bit (LSB) of the quantized DCT coefficients with the secret messages. However, the message capacity of JPEG-JSTEG is very limited. In order to increase the capacity of the secret message, Chang et al. [13] proposed a Joint Quantization Table Modification (JQTM) method based on the modification of the standard JPEG quantization table and the secret messages are embedded in the middle frequency of DCT coefficients of each 8×8 block. The hiding capacity of the JQTM method is significantly more as compared to the JPEG-JSTEG; however, there are fewer (limited to 26) quantized DCT coefficients for each block to hide the secret message. Another limitation of this method is having a low-security level of secret message [40]. Thus, in order to work on the said limitations, researchers [33, 47, 64, 79] proposed few methods where the concept of metaheuristic optimization is introduced and some of them worked on the standard quantization table used for JPEG image. The following section discusses various practical steganography applications.

2.3 Steganography Applications

There are numerous applications of image steganography [75]. One of the most common applications is watermarking. A digital watermark can be inserted into an image to provide security to the image. A secret chemical formula can be hidden in a cover image using steganography for a new invention. Another application of steganography could be the extension of controlling access rights for unauthorized users while distributing digital content by issuing one secret key [32]. Furthermore, annotation data or metadata could be easily added in a media database system which could further help to implement a keyword-based movie scene retrieving system efficiently [53]. Although there are several applications of image steganography, the major concern identified by the researchers is the counterpart of steganography referred to as steganalysis [78]. The motive of using image steganalysis [54, 58] is to reduce the efficiency of image steganography and to extract the secret hidden text which further creates a need to have a secure steganography method.

2.4   Summary

This chapter discussed the overview of cryptography and digital steganography. The important techniques of cryptography and steganography are explained. The detailed review of JPEG image compression and its importance are also discussed. The underlying concept using different quantization tables having different sizes is also briefed. In order to increase the secret text embedding capacity, the size of the standard quantization table could be increased. Furthermore, the importance of steganography and its applications are explored. The reverse of image steganography referred to as image steganalysis and the importance of developing a secure steganography method are also discussed. The next chapter analyses several heuristic and metaheuristic optimization algorithms.

References

1. Acharya, B., Shukla, S.K., Panigrahy, S.K., Patra, S.K., Panda, G.: H-S-X cryptosystem and its application to image encryption. In: International Conference on Advances in Computing, Control, and Telecommunication Technologies, 28–29 Dec 2009, IEEE, Trivandrum, Kerala, India (2009)
2. Ahmad, M., Khan, I.R., Alam, S.: Cryptanalysis of Image encryption algorithm based on fractional-order lorenz-like chaotic system. In: Satapathy, S.C. et al. (eds.) Emerging ICT for Bridging the Future, Advances in Intelligent Systems and Computing, Proceedings of the 49th Annual Convention of the Computer Society of India CSI, vol. 2, pp 381–388 (2015)
3. Akritas, A.G., Iyengar, S.S., Rampuria, A.A.: Computationally efficient algorithms for a one-time pad scheme. Int. J. Comput. Inform. Sci. **12**(4), 285–316 (1983)
4. Aljohani, M., Ahmad, I., Basheri, M., Alassafi, M.O.: Performance analysis of cryptographic pseudorandom number generators special section on roadmap to 5G: rising to the challenge. IEEE Access **7**, 39794–39805 (2019)
5. Arab, A., Rostami, M.J., Ghavami, B.: An image encryption method based on chaos system and AES algorithm. J. Supercomput. **75**(10), 6663–6682 (2019)
6. Arham, A., Nugroho, H.A., Adji, T.B.: Multiple layer data hiding scheme based on difference expansion of quad. Signal Process. **137**, 52–62 (2017)
7. Bhaskaran, V., Konstantinides, K.: Methods and standards for lossless compression. In: Image and Video Compression Standards, The Springer International Series in Engineering and Computer Science, vol. 334, pp. 15–51. Springer, Boston, MA (1995)

8. Bhattacharyya, D., Kim, T.: Image data hiding technique using discrete fourier transformation. In: Kim, T., Adeli, H., Robles, R.J., Balitanas, M. (eds.) Ubiquitous Computing and Multimedia Applications (UCMA), Communications in Computer and Information Science, vol. 151, pp. 315–323. Springer, Berlin (2011)

9. Biryukov, A., Kushilevitz, E.: From differential cryptanalysis to ciphertext-only attacks. In: Krawczyk, H. (ed.) Advances in Cryptology—CRYPTO '98, Lecture Notes in Computer Science (LNCS), vol. 1462, pp. 72–88. Springer, Berlin (1998)

10. Blake, I.F., Garefalakis, T.: On the security of the digital signature algorithm. Des. Codes Crypt. **26**(1–3), 87–96 (2002)

11. Bohme, R.: Principles of modern steganography and steganalysis. Advanced Statistical Steganalysis, Information Security and Cryptography, vol. 0, pp. 11–77 (2010)

12. Chan, C.K., Cheng, L.M.: Hiding data in images by simple LSB substitution. Pattern Recogn. **37**(3), 469–474 (2004)

13. Chang, C.C., Chen, T.S., Chung, L.Z.: A steganographic method based upon JPEG and quantization table modification. Inf. Sci. **141**, 123–138 (2002)

14. Cheddad, A., Condell, J., Curran, K., Kevitt, P.: Digital Image Steganography: Survey and Analysis of Current Methods. Sig. Process. **90**(3), 727–752 (2010)

15. Chen, J., Xue, D., Lai, X.: An analysis of international data encryption algorithm (IDEA) security against differential cryptanalysis. Wuhan Univ. J. Nat. Sci. **13**(6), 697–701 (2008)

16. Clavier, C., Feix, B., Gagnerot, B., Roussellet, M.: Passive and active combined attacks on AES combining fault attacks and side channel analysis. In: 2010 Workshop on Fault Diagnosis and Tolerance in Cryptography, pp. 10–19. Santa Barbara, CA (2010)

17. Coron, J.S.: What is cryptography? IEEE Secur. Priv. **4**(1), 70–73 (2006)

18. Debra, L.C., Yung, M., Keromytis, A.D.: Methods for linear and differential cryptanalysis of elastic block ciphers. In: Mu, Y., Susilo, W., Seberry, J. (eds.) Information Security and Privacy, ACISP, Lecture Notes in Computer Science (LNCS), vol. 5107, pp. 187–202. Springer, Berlin (2008)

19. Desmedt, Y., Odlyzko, A.M.: A chosen text attack on the RSA cryptosystem and some discrete logarithm schemes. In: Proceedings of Lecture notes in computer sciences. Advances in cryptology—CRYPTO, 85 Mar 2007, vol. 218, pp. 516–522 (2007)

20. Diffie, W., Hellman, M.E.: Exhaustive cryptanalysis of the NBS data encryption standard. IEEE Comput. **10**(6), 74–84 (1977)

21. Dorigo, M., Birattari, M., Stitzle, T.: Ant colony optimization: artificial ants as a computational intelligence technique. IEEE Comput. Intell. Mag. **1**(4), 28–39 (2006)

22. Douglas, M., Bailey, K., Leeney, M., Curran, K.: An overview of steganography techniques applied to the protection of biometric data. Multimed. Tools Appl. **77**(13), 17333–17373 (2018)

23. Duraisamy, J.H., Subramaniyan, U., Daniela, E.P., Antoanela, N.: Application of genetic algorithm and particle swarm optimization techniques for improved image steganography systems. Open Phys. **14**(1), 452–462 (2016)

24. Edwina Alias, T., Mathew, D., Thomas, A.: Steganographic technique using secure adaptive pixel pair matching for embedding multiple data types in images. In: Proceedings of the 5th International Conference on Advances in Computing and Communications, ICACC 2015, pp. 426–429. Sept 2015

25. Gaj, K., Orlowski, A.: Facts and myths of enigma: breaking stereotypes. In: EUROCRYPT 2003, International Association for Cryptologic Research, LNCS, vol. 2656, pp. 106–122 (2003)

26. Gardiner, J.: Stegchat: a synonym-substitution based algorithm for text steganography. Submitted in conformity with the requirements for the degree of MSc Computer Security, School of Computer Science, University of Birmingham, pp. 1–63 (2012)

27. Ghasemi, E., Shanbehzadeh, J., Fassihi, N.: High capacity image steganography based on genetic algorithm and wavelet transform. In: Ao, S., Castillo, O., Huang, X. (eds.) Intelligent Control and Innovative Computing. Lecture Notes in Electrical Engineering, vol. 110, pp. 395–404. Springer, Boston, MA (2012)

28. Giraud, C., Thiebeauld, H.: A survey on fault attacks. In: Quisquater, J.J., Paradinas, P., Deswarte, Y., El Kalam, A.A. (eds.) Smart Card Research and Advanced Applications VI, IFIP International Federation for Information Processing, vol. 153, pp. 159–176. Springer, Boston, MA (2004)
29. Gupta, A., Dhami, A.: Measuring the impact of security, trust and privacy in information sharing: a study on social networking sites. J. Direct Data Digit. Mark. Pract. **17**(1), 43–53 (2015)
30. Heys, H.M., Tavares, S.E.: Known plaintext cryptanalysis of tree-structured block ciphers. Electron. Lett. **31**(10), 784–785 (1995)
31. Johansson, T.: Rabbit: a new high-performance stream cipher. In: Johansson, T. (ed.) Fast Software Encryption, FSE 2003, Lecture Notes in Computer Science (LNCS), vol. 2887, pp. 307–329. Springer, Berlin (2003)
32. Kawaguchi, E., Maeta, M., Noda, H., Nozaki, K.: A model of digital contents access control system using steganographic information hiding scheme. In: Proceedings of Information Modeling and Knowledge Bases XVIII, pp. 50–61. Trojanovice, Czech Republic, 29 May–2 June 2006 (2006)
33. Khamrui, A., Mandal, J.K.: A genetic algorithm based steganography using discrete cosine transformation (GASDCT). In: Proceedings of International Conference on Computational Intelligence: Modeling Techniques and Applications (CIMTA), pp. 105–111. University of Kalyani, West Bengal, India, 27–28 Sept 2013 (2013)
34. Klapetek, P., February 2002. http://klapetek.cz/wdwt.html. Last accessed on 06.11.19
35. Kobayashi, H., Noguchi, Y., Kiya, H.: A method of embedding binary data into JPEG bit streams. IEICE Trans. Inf. Syst. **83**-D-II, 1469–1476 (1999)
36. Krishnamurthy, G.N., Ramaswamy, V., Leela, G.H.: Performance enhancement of blow-fish algorithm by modifying its function. In: Sobh, T., Elleithy, K., Mahmood, A., Karim, M. (eds.) Innovative Algorithms and Techniques in Automation, Industrial Electronics and Telecommunications, pp. 241–244. Springer, Dordrecht (2007)
37. Kumar, S., Singh, A., Kumar, M.: Information hiding with adaptive steganography based on novel fuzzy edge identification. Def. Technol. **15**(2), 162–169 (2019)
38. Kumar, A., Mehra, P.S., Gupta, G., Sharma, M.: International conference on heterogeneous networking for quality, reliability, security and robustness. In: Quality, Reliability, Security and Robustness in Heterogeneous Networks, Lecture Notes of the Institute for Computer Sciences, Social Informatics and Telecommunications Engineering (LNICST), vol. 115, pp. 689–695. (2013)
39. Kumar, S., Wollinger, T.: Fundamentals of symmetric cryptography. In: Lemke, K., Paar, C., Wolf, M. (eds.) Embedded Security in Cars, pp. 125–143. Springer, Berlin (2006)
40. Li, X., Wang, J.: A steganographic method based upon JPEG and particle swarm optimization algorithm. Inf. Sci. **177**, 3099–3109 (2007)
41. Li, M., Guo, Y., Wang, B., Kong, X.: Secure spread-spectrum data embedding with PN-sequence masking. In: Signal Processing: Image Communication, vol. 39(Part A), pp. 17–25 (2015)
42. Li, Z.N., Drew, M.S., Liu, J.: Lossy compression algorithms. In: Fundamentals of Multimedia: Texts in Computer Science, pp. 225–280. Springer, Berlin (2014)
43. Lorch, B., Riess, C.: Image forensics from chroma subsampling of high-quality JPEG images. In: IH&MMSec'19 Proceedings of the ACM Workshop on Information Hiding and Multimedia Security, pp. 101–106. 03–05 July 2019 (2019)
44. Malik, A., Sikka, G., Verma, K.H.: A modified pixel-value differencing image steganographic scheme with least significant bit substitution method. Int. J. Image Graph. Signal Process. (IJIGSP) **4**(4), 68–74 (2015)
45. Maniriho, P., Ahmad, T.: Information hiding scheme for digital images using difference expansion and modulus function. J. King Saud Univ. Comput. Inf. Sci. **31**(3), 335–347 (2019)
46. Mogos, G.: Quantum key distribution protocol with four-state systems—software implementation. Procedia Comput. Sci. **54**, 65–72 (2015)
47. Nickfarjam, A.M., Azimifar, Z.: Image steganography based on pixel ranking and particle swarm optimization. In: Proceedings of the 16th CSI International Symposium on Artificial Intelligence and Signal Processing (AISP 2012), IEEE, Shiraz, Fars, Iran, 2–3 May 2012 (2012)

48. Noda, H., Furuta, T., Niimi, M., Kawaguchi, E.: Application of BPCS steganography to wavelet compressed video. In: Proceedings of International Conference on Image Processing, vol. 4, pp. 2147–2150. November 2004 (2004)
49. Paar, C., Pelzl, J.: The Advanced Encryption Standard (AES). In: Understanding cryptography, pp. 87–121. Springer, Berlin (2010)
50. Pachghare, V.K.: Cryptography and Information Security, p. 416. PHI Learning Private Limited, New Delhi (2015)
51. Pak, C., Kim, J., An, K., Kim, C., Kim, K., Pak, C.: A novel color image LSB steganography using improved 1D chaotic map. Multimed. Tools Appl. 1–17 (2019)
52. Pannebaker, W.B., Mitchell, L.: JPEG: still image data compression standard. Van Nostrand Reinhold, vol. 638, New York (1993)
53. Patel, N.S., Abowd, D.G.: The contextcam: automated point of capture video annotation. In: Davies, N., Mynatt, E.D., Siio, I. (eds.) UbiComp 2004: Ubiquitous Computing. UbiComp 2004, Lecture Notes in Computer Science (LNCS), vol. 3205, pp. 301–318. Springer, Berlin (2004)
54. Pathak, S., Roy, R., Changder, S.: Performance analysis of image steganalysis techniques and future research directives. Int. J. Inf. Comput. Secur. (IJICS) 10(1), 1–24 (2018)
55. Patil, R.B., Kulat, K.D.: Image and text compression using dynamic huffman and RLE coding. In: Deep, K., Nagar, A., Pant, M., Bansal, J. (eds.) Proceedings of the International Conference on Soft Computing for Problem Solving (SocProS 2011), December 20–22, 2011, Advances in Intelligent and Soft Computing, vol. 131, pp. 701–708. Springer, New Delhi (2011)
56. Petitcolas, F.A.P., Anderson, R.J., Kuhn, M.G.: Information hiding-a survey. In: Proceedings of the IEEE, Special Issue on the Protection of Multimedia Content, vol. 87, no. 7, pp. 1062–1078 (1999)
57. Preneel, B.: Cryptographic hash functions: theory and practice. In: Gong, G., Gupta, K.C. (eds.) Progress in Cryptology—INDOCRYPT 2010. Lecture Notes in Computer Science, vol. 6498, pp. 115–117. Springer, Berlin (2010)
58. Qiao, X., Ji, G., Zheng, H.: A new method of steganalysis based on image entropy. In: Huang, D.S., Heutte, L., Loog, M. (eds.) Advanced Intelligent Computing Theories and Applications, With Aspects of Contemporary Intelligent Computing Techniques. ICIC 2007, Communications in Computer and Information Science, vol. 2, pp. 810–815. Springer, Berlin (2007)
59. Rabah, K.: Steganography-the art of hiding data. Inf. Technol. J. 3(3), 245–269 (2004)
60. Raid, A.M., Khedr, W.M., El-dosuky, M.A., Ahmed, W.: JPEG image compression using discrete cosine transform—a survey. Int. J. Comput. Sci. Eng. Surv. (IJCSES) 5(2), 60–70 (2014)
61. Rao, K., Yip, P.: Discrete Cosine Transform: Algorithms, Advantages, Applications. Academic Press Professional, Inc. San Diego, CA (1990) ISBN: 0-12-580203-X
62. Rao, U.H., Nayak, U.: Cryptography. The InfoSec Handbook, pp. 163–181. Apress, Berkeley, CA (2014)
63. Rivest, R.: The MD5 Message-Digest Algorithm. ACM Digital Library, RFC, New York (1992)
64. Roy, R., Laha, S.: Optimization of stego image retaining secret information using genetic algorithm with 8-connected PSNR. In: Proceedings of 19th International Conference on Knowledge Based and Intelligent Information and Engineering Systems, pp. 468–477. 7–9 Sept 2015 (2015)
65. Salomon, D.: Data Privacy and Security (2003) Springer, Berlin. ISBN: 978-0-387-21707-9
66. Sarmah, D., Kulkarni, A.J.: JPEG based steganography methods using cohort intelligence with cognitive computing and modified multi random start local search optimization algorithms. Inf. Sci. 430–431, 378–396 (2018)
67. Sarmah, D.K., Kulkarni, A.J.: Improved cohort intelligence-a high capacity, swift and secure approach on JPEG image steganography. J. Inf. Secur. Appl. 45, 90–106 (2019)
68. Sharma, K., Gupta, K.: Lossless data compression techniques and their performance. In: Proceedings of International Conference on Computing, Communication and Automation (ICCCA), pp. 256–261. IEEE, Greater Noida, India, 5–6 May 2017 (2017)

69. Shivani Yadav, V.K., Batham, S.: A novel approach of bulk data hiding using text steganography. Procedia Comput. Sci. **57**, 1401–1410 (2015)
70. Spezeski, W.J.: A keyless polyalphabetic cipher. Technological Developments in Networking, Education and Automation, pp. 529–532. Springer, Berlin (2010)
71. Stallings, W.: The advanced encryption standard. Cryptologia **26**(3), 165–188 (2002)
72. Sun, S.: A novel edge based image steganography with 2^k correction and huffman encoding. Inf. Process. Lett. **116**(2), 93–99 (2016)
73. Tutun, S., Akca, M., Bıyıklı, O., Khasawneh, M.T.: An outlier-based intention detection for discovering terrorist strategies. Procedia Comput. Sci. **114**, 132–138 (2017)
74. Upham, D., JPEG-Jsteg-v4. http://www.funet.fi/pub/crypt/steganography/jpeg-jsteg-v4.diff. gz. Last accessed on 6.11.19
75. Venkatraman, S., Abraham, A., Paprzycki, M.: Significance of steganography on data security. In Proceedings of International Conference on Information Technology: Coding and Computing, Proceedings ITCC, pp. 1–5. 5–7 Apr 2004 (2004)
76. Verma, J., Prasad, S.: Security enhancement in data encryption standard. In: Prasad, S.K., Routray, S., Khurana, R., Sahni, S. (eds.) Information Systems, Technology and Management, ICISTM 2009. Communications in Computer and Information Science, vol. 31, pp. 325–334. Springer, Berlin (2009)
77. Wang, R.Z., Lin, C.F., Lin, J.C.: Image hiding by optimal LSB substitution and genetic algorithm. Pattern Recogn. **34**(3), 671–683 (2001)
78. Wang, H., Wang, S.: Cyber warfare: steganography vs. steganalysis. Commun. ACM Voting Syst. **47**(10), 76–82 (2004)
79. Wang, S., Yang, B., Niu, X.: A secure steganography method based on genetic algorithm. J. Inf. Hiding Multimed. Signal Process. **1**(1), 28–35 (2010)
80. Wardlaw, W.P.: The RSA public key cryptosystem. In: Proceedings of Coding Theory and Cryptography, pp. 101–123. Springer, Berlin (2000)
81. Wasan, S.A.: Information hiding using ant colony optimization algorithm. Int. J. Technol. Diffus. **2**(1), 16–28 (2011)
82. Westfeld, A., Pfitzmann, A.: Attacks on steganographic systems breaking the steganographic utilities ezstego, jsteg, steganos, and S-tools-and some lessons learned. In: Information Hiding, Third International Workshop, IH'99, 1 Lecture Notes in Computer Science (LNCS), pp. 61–76. Springer, Dresden, Germany (1999)
83. Wu, H., Preneel, B.: Differential-linear attacks against the stream cipher phelix. In: Biryukov, A. (ed.) Fast Software Encryption (FSE), Lecture Notes in Computer Science (LNCS), vol. 4593, pp. 87–100. Springer, Berlin (2007)
84. Yan, S.Y.: Quantum Attacks on Public-Key Cryptosystems. Springer, Berlin, US (2013) ISBN: 978-1-4419-7722-9 (eBook)
85. Yoon, J.W., Kim, H., Jo, H., Lee, H., Lee, K.: Visual honey encryption: application to steganography. In: IH&MMSec '15, Proceedings of the 3rd ACM Workshop on Information Hiding and Multimedia Security, pp. 65–74. 17–19 June 2015 (2015)
86. Zhang, X., Wang, S.: Steganography using multiple-base notational system and human vision sensitivity. IEEE Signal Process. Lett. **12**(1), 67–70 (2005)
87. Zhang, X., Wang, S.: Efficient steganographic embedding by exploiting modification direction. IEEE Commun. Lett. **10**(11), 781–783 (2006)
88. Zhang, W., Wang, S., Zhang, X.: Improving embedding efficiency of covering codes for applications in steganography. IEEE Commun. Lett. **11**(8), 680–682 (2007)
89. Zhang, C., Zhang, Y.: Authenticated diffie-hellman key agreement protocol with forward secrecy. Wuhan Univ. J. Nat. Sci. **13**(6), 641–644 (2008)
90. Zou, Y., Zhang, G., Liu, L.: Research on image steganography analysis based on deep learning. J. Vis. Commun. Image Represent. **60**, 266–275 (2019)

Chapter 3
Heuristics and Metaheuristic Optimization Algorithms

The concept of optimization is widely used to locate the optimum solution for different mathematical problems. A mathematical model is prepared to solve real-world problems where various constraints are inherently associated with it. Heuristics [15] and Metaheuristics [22] are the search techniques for solving various operational research and logistics problems. These techniques are efficient than classical methods such as enumerative [14], branch and bound [31], dynamic programming [25], simplex algorithm [73], linear and integer programming [54], etc. Also, they help in solving complex real-world problems in an efficient way where the large dimensions, higher constraints and time factor [65] are involved in the problems. The Heuristic algorithms do not guarantee optimality, accuracy, precision, or speed. However, they provide good solutions and can be verified when required. One can employ the Heuristic algorithms when finding the exact solution is not a criterion; the approximate solution is enough for the problem. In this algorithm, a Heuristic function keeps a check at each branching step of search algorithms to evaluate the information which enables us to recognize which branch to track. On the other hand, the advanced form of heuristics is referred to as meta-heuristics. The Heuristics algorithms are problem-specific whereas Metaheuristics are problem independent and generic to solve a large variety of real-world problems. The metaheuristic algorithms also do not guarantee to find the optimal solution to a given problem. In the next sections, the Heuristics and metaheuristics are explained with the help of examples.

3.1 Heuristic Algorithms with Examples

The Heuristics search techniques are classified into two categories: Direct [44] and Weak [44] as shown in Fig. 3.1a. The Direct search technique referred to as Blind or Uninformed search, generally requires the entire solution space for search. It also

D. K. Sarmah et al., *Optimization Models in Steganography Using Metaheuristics*, Intelligent Systems Reference Library 187, https://doi.org/10.1007/978-3-030-42044-4_3

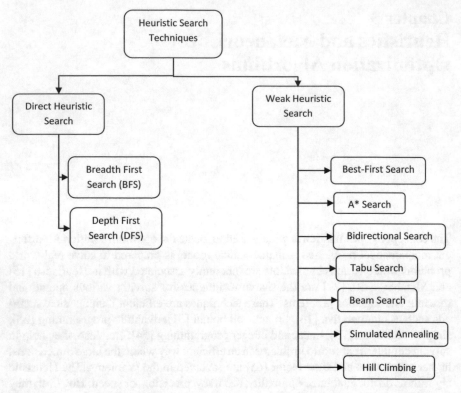

Fig. 3.1 Heuristics search techniques

takes a large time and consumes more memory space to find the solution. Breadth-First Search (BFS) [71] and Depth First Search (DFS) [71] are the two methods under this technique. The Weak heuristic search technique, the second category of Heuristic algorithms is an effective technique if necessary domain-specific information is provided and applied properly based on the application. This technique is also referred to as Informed Search, which has several algorithms as portrayed in Fig. 3.1a: Best–First Search [68], A* Search [19], Bidirectional Search [18], Tabu Search [5], Beam Search [55], Simulated Annealing [12], Hill Climbing [7], etc.

There are several kinds of problems which could be solved by Heuristic techniques. A few of the well-known problem examples are Travelling Salesman Problem [46], Knapsack Problem [37], Virus Scanning [2], Searching and Sorting [74], Maze problem [57], Towers of Hanoi [6], etc. Based on their solving strategy, the Heuristic technique is divided into two schemes i.e. construction heuristics and descending heuristics. The construction heuristics identify the solutions via iterations, for example, Greedy methods [3]. However, the descending heuristics search for local optima solution from available solutions. The next section elaborates on the metaheuristic algorithms with examples.

3.2 Metaheuristic Algorithms with Examples

The advanced form of Heuristics is referred to as Metaheuristics where researchers focused on developing certain generalized optimization techniques inspired from biological and natural instincts to solve such real-world problems. Metaheuristic algorithms can be applied to a variety of problems to seek a near-optimal solution. There are several metaheuristic optimization algorithms developed so far. Evolutionary Algorithms (EAs) [13] fall under the category of bio-inspired optimization algorithms. Such algorithms are inspired by the processes of biological systems. The popular algorithms under this category are the Genetic Algorithm (GA) [26, 60, 4], Differential Evolution (DA) [20], etc. Closer to the bio-inspired algorithms, Swarm Intelligence (SI) [8] is inspired from the collective behavior exhibited by a number of living species such as insects, fishes, ants, bees, etc. The concept of SI is based on the cooperative behavior of individuals in a certain direction to achieve success on the shared goal. The popular examples under this category are Particle Swarm Optimization (PSO) [21], Ant Colony Optimization (ACO) [10, 11], Bees Algorithm (BA) [81], Firefly Algorithm (FA) [79], Bat Algorithm [80], etc. Another successful alternative of EA has been recognized as Socio-Inspired Optimization (SIO) algorithms [72]. Such algorithms are motivated by the social behavior of a human being. In SIO, every individual interacts with one another to improve its overall behavior, which may further increase the possibility of achieving a certain goal in the entire society. The optimization algorithms in this category are Cohort Intelligence (CI), Teaching-Learning Based Optimization (TLBO) [59], Social Group Optimization (SGO) [66], Social Learning Optimization (SLO) [48], Cultural Evolution Algorithm (CEA) [43], Social Emotional Optimization (SEO) [78], League Championship Algorithm (LCA) [30], etc.

This chapter discusses the different categories of metaheuristic algorithms. A detailed review of socio inspired optimization algorithms and Cohort Intelligence (CI) [39, 63] is described in Sect. 3.2.1 and 3.2.2 respectively. Further, the concept of steganography used for information security with various optimization algorithms has also been discussed.

3.2.1 Review of Socio-Inspired Optimization (SIO) Algorithms

Nowadays SIO algorithms are quite popular because of their approach to locate the optimized solution for complex problems. The behavior of such algorithms is inspired by observing individual learning from one another through interactions in society. In these algorithms, individual(s) imitates others from within the society to improve their social characteristics. Numerous algorithms were developed under this category in which few of them are listed above. In 1994, one of the SIO algorithms developed by Reynolds is referred to as Cultural Algorithm. This algorithm is based on the principle of cultural evolution where culture is represented as behavior, knowledge, ethics, manners, etc. of an individual of the society. This algorithm suggests that individual evolves much faster through cultural evolution than though biological or genetic evolution. On similar lines, many researchers had put up their efforts to propose and develop their own mathematical model under SIO. Few latest SIO algorithms are: Society and Civilization Optimization (SCO) algorithm [67], Imperialist Competitive Algorithm (ICA) [16], League Championship algorithm (LCA) [30], Social Emotional Optimization (SEO) [78], Election Campaign Optimization (ECO) algorithm [49], Anarchic Society Optimization (ASO) algorithm [27], Teaching–learning-based optimization (TLBO) [59], Cohort Intelligence (CI) [39], Soccer League Competition (SLC) algorithm [51], Social Learning optimization (SLO) [48], Social Group Optimization (SGO) [66], Ideology Algorithm (IA), Socio Evolution & Learning Optimization (SELO) algorithm [42, 41], Multi-cohort Intelligence Algorithm (MCIA) [70], etc. The idea to develop SCO [58] was originated by observing the behavior of social interactions which enable individuals to adapt and advance their culture within a society. Socio-political behaviors between the imperialist nations were observed in ICA [16, 22]. The individuals of the imperialist nations compete with one another. The powerful one takes possession of other's colonies and forms its own empire. This algorithm was used to solve different types of optimization problems.

LCA is another popular algorithm under the SIO category proposed by Kashan [30]. This algorithm was developed for global optimization and inspired by sports championships seen in league matches. The Sports teams having a number of individuals competing for multiple weeks (iterations) as per a league schedule. In the end, the outcome of the game was determined by declaring the winner, emerged gradually from the strong team with higher fitness value. SEO algorithm [78] is a swarm-based technique that simulates human behavior. This algorithm was conceptualized by considering an individual in a society who wanted to attain a superior position. Decisions taken by the individual were based on their current behavior and controlled by the corresponding emotional index. In order to increase their personal status, they interacted and competed with other individuals in society. Finally, society selected the individual based on their behavior. ECO algorithm [49] is based on the behavior of election candidates seek maximum support from voters during a campaign. As the number of inspired voters increased the number of supporters was also increased for them. The prestige of the candidate during the campaign decided the number of supporters for them. During experiments, this algorithm was tested with 3 optimization functions and the PID controller parameters tuning problem. The other efficient algorithm i.e. ASO algorithm [27] motivated by a group of individuals whose behavior was negative towards the environment, fickle, irrational, etc. Due to the anarchic behaviors of the members, ASO explored the entire search space and locate better solutions. TLBO [59] is another popular socio inspired optimization algorithm that seeks inspiration from the teaching-learning methodology of a classroom. This algorithm was partitioned into two phases: the first phase was referred to as the 'Teacher phase' and was used to gain knowledge from the teacher. The other phase was referred to as the 'Learner phase' wherein the learner improved their knowledge by interacting with each other. As proposed and developed by Kulkarni et al. [39], an emerging metaheuristic optimization algorithm referred to as CI modeled the self-learning behavior of candidates in a cohort. These candidates participated, interacted and competed with one another to achieve one common goal. Soccer League matches were the inspiration for Moosavian and Roodsari [52] to develop the SLC algorithm. It was based on the competition between teams and players. This algorithm was tested and validated with respect to the parameters such as search accuracy, reliability and convergence speed. SLO algorithm [48] is based on the social learning theory which seeks inspiration from human intelligence evolution and development. The latest population-based optimization algorithm SGO, developed by Satapathy and Naik [66] was motivated by human behavior to solve a complex problem. The validation of this algorithm was done by applying this algorithm to solve various unconstrained benchmark functions and standard numerical benchmark functions. IA [23] presented an idea towards political party individuals who competed with other individuals in the same political party or leaders of the other political party to achieve their goal.

The emerging and very recent metaheuristic optimization algorithm SELO was proposed by Kumar et al. [41]. This algorithm worked towards adapting the behavior and learning from members of the same family and different families in society to achieve one common goal. The performance was validated by solving 50 test functions. The results of this algorithm were satisfactory in some cases and comparable with the other popular algorithms. Recently an upcoming algorithm, MCIA was developed by Shastri and Kulkarni [70], based on interaction amongst different cohorts. The performance of this algorithm was successfully tested experimentally by solving 75 unconstrained test problems having dimensions up to 30.

In the proposed research, CI [39, 63, 64] and MRSLS optimization algorithms [35, 63, 64] were extensively explored and used in the area of information hiding. The following section discusses the CI optimization algorithm in detail.

3.2.2 Cohort Intelligence (CI) Optimization Algorithm

A socio-inspired optimization algorithm referred to as CI was developed by Kulkarni et al. [39]. In the framework of CI [38], every candidate observes its own behavior and tries to improve it using interaction and competition. In each learning attempt, every candidate chooses a certain candidate in the cohort from which it can learn certain qualities. This may make the follower candidate improve its current behavior. The quality here refers to the variables in the system and the behavior refers to the objective function. The details of this algorithm are presented below: The number of candidates is assumed C in the cohort and every individual candidate c $(c = 1, 2, \ldots, C)$ belongs to a set of qualities $x^c = \left(x_1^c, x_2^c, x_3^c, \ldots, x_N^c \right)$. The behavior of the candidate depends on its overall quality and can be expressed as $f(x^c) = f\left(x_1^c, x_2^c, x_3^c, \ldots, x_N^c \right)$. The individual behavior of each candidate c is being observed by itself and every other candidate c' in the cohort. Each candidate c may try to improve its current behavior $f^*(x^c)$ by following the better behavior $f^*(x^{c'})$ in the cohort: $f^*(x^c) < f^*(x^{c'})$. This helps to evolve the entire behavior of the cohort. Since certain candidate to select better behavior from fellow candidates based on probability, there is always uncertainty associated with it. The entire cohort behavior improves until saturation where the candidates could not see any further improvement in its behavior. The procedure begins with the following steps:

Step 1: Initialize the number of candidates C where any candidate is represented as $c = 1, 2, \ldots, C$, learning attempt counter $n = 1$ and the convergence factor r. The value of C and r can be chosen based on the preliminary trials of the algorithm.

Step 2: Calculate the probability of each candidate by selecting the behavior of every other associated candidate, where $f^*(x_c)$ refers to the behavior of the individual candidate.

$$p^c = \frac{\frac{1}{f^*(x_c)}}{\sum_{c=1}^{C} \frac{1}{f^*(x^c)}} \tag{3.1}$$

Step 3: A random number $r \in (0, 1)$ is generated by every candidate and based on a roulette wheel selection approach a candidate is chosen whose behavior could be followed.

Step 4: Saturation/convergence condition is evaluated if the difference between the individual behaviors of the candidates is not very substantial for a successive considerable number of learning attempts, where l is considered as a learning attempt i.e.

$$|\max(f(x^c)^l) - \max(f(x^c)^{l-1})| \le \varepsilon, \text{ and} \tag{3.2}$$

$$|\min(f(x^c)^l) - \min(f(x^c)^{l-1})| \le \varepsilon, \text{ and} \tag{3.3}$$

$$|\max(f(x^c)^l) - \min(f(x^c)^l)| \le \varepsilon. \tag{3.4}$$

Step 6: Accept the candidate's behavior as the final solution, if the following two conditions are satisfied, else continue to step 2.

(a) The condition mentioned in Step 4.
(b) If the maximum number of learning attempts exceeded.

The CI algorithm was validated by solving several unconstrained test problems [38, 39]. The algorithm then applied for solving NP-hard problems such as 0-1 knapsack problems [38, 37], health care and supply chain management problems (Kulkarni et al. 2016; [38], Travelling Salesman Problem [38], Sea Cargo Mix problem and selection of Cross Border Shipper problem [35]. The modified CI (MCI) along with its hybridized version with K-means (K-MCI) was proposed and applied for solving data clustering problems [34, 38]. The algorithm of CI was also applied for optimizing the manufacturing cost for the shell-and-tube heat exchanger [9].

Furthermore, for solving constrained metaheuristic problems using CI, two methods were proposed by [40]. These methods were (i) CI with static penalty function and (ii) CI with dynamic penalty function. The validation of these methods was done by analyzing and comparing with the algorithms such as GA, PSO, ABC, d-Ds. The performance of CI was found better for solving several constrained problems compared to contemporary algorithms. The CI algorithms were successfully applied for solving discrete and mixed variable problems from the truss design and mechanical engineering domain [28]. The algorithm performed better as compared to Probability Collectives (PC) [36, 39], PSO [45], and GA [17, 77, 76]. Furthermore, Shah et al., [69] applied the CI algorithm to optimize the parameters of fractional PID controller. The performance was validated by comparing it with GA and PSO algorithms. Recently, seven variations of CI were proposed by Patankar and Kulkarni [56] solving seven multimodal and three uni-modal unconstrained test functions. The CI algorithm performed exceedingly better in solving the problems from the continuous domain, combinatorial domain, and several unconstrained and constrained test problems, etc. This algorithm was further investigated to improve mesh quality by Sapre et al. [62]. Furthermore, the CI algorithm was applied and validated to solve different types of engineering problems [29, 50]. In addition to this, CI was employed in the domain of swarm robotics [61].

Along with advantages, few limitations of CI were also observed. The computational performance of this algorithm was controlled by the parameters such as a number of candidates and sampling interval reduction factors which could be judged by the quality of solution and rate of convergence. Thus, there is a requirement to develop a self-adaptive system for CI which could fine-tune these parameters on its own. Further, the algorithm could be improved to solve more variety of application-oriented and real-world problems.

The next section presents the state of art for steganography using various optimization algorithms.

3.2.3 Multi Random Start Local Search (MRSLS) Optimization Algorithm

MRSLS optimization method is proposed by [35] to solve the constrained combinatorial problems. The MRSLS is a technique based on the pairwise solution interchange approach. In order to search for better solutions, the elements of two adjacent solutions are randomly swapped. The details of the original MRSLS methodology are presented below.

The steps for the algorithm are as follows:

Step 1: Consider a set $s = \{1, 2, 3, \ldots, n\}$ for which an optimal solution needs to be determined.

Step 2: Assume a starting solution is randomly generated and value is achieved.

Step 3: Use a pair-wise exchange approach to interchange neighboring solutions. The neighbors are considered as the two elements occupying the adjacent positions.

Step 4: Calculate and update the current solution if it is found superior to the previously calculated solutions.

Step 5: Continue the process of evaluation until the stopping criterion is met.

This method has been tested and validated with CI for small scale, medium scale, and large scale test problems. This method provides a very good and comparable result to solve the NP-hard problem such as the Cyclic Bottleneck Assignment Problem (CBAP) [35]. As the initial solution is randomly generated in MRSLS, sometimes these random solutions could be infeasible and may require to construct the initial feasible solutions.

3.3 Steganography Using Existing Optimization Algorithms

As mentioned in Sect. 3.2, the image steganography is used to hide secret data in the cover image. The resultant image with embedded secret data is referred to as the stego image. As an important field in information security, steganography gained popularity amongst researchers [47, 75]. Various steganography algorithms have been developed to enhance the associated important parameters such as image quality, secret text embedding capacity and secret text security. The detailed analysis of such algorithms is presented in Sect. 3.2. The notable metaheuristic optimization algorithms have been combined with image steganography by observing the limitations of the previously developed steganography algorithms. Li and Wang [47] proposed and developed a PSO [32] based JPEG image steganography algorithm referred to as JPEG_PSO. In this approach, the quantization table used in JQTM was modified to increase the capacity of a secret message. The process of modifying the standard quantization table was taken from Huang et al. [24]. Further, Li and Wang [47] proposed the approach of embedding the secret message into low to middle-frequency components for each block. This approach was inspired by the Optimal LSB Substitution (OLSBS) approach [75] in which the secret message was first transformed into a ciphertext with the help of an optimal substitution matrix and then the transformed results were embedded into the cover image. The substitution matrices were considered as the variation of identity matrices. Wang et al. [75] employed GA [33, 60] to search the optimal substitution matrix which improved the stego image quality in the spatial domain. PSO (Li and Wang [47] was used to select the optimal substitution matrix where a particle having 2^K dimensions was described by a substitution matrix. The value of K decided the number of the group of bit value to be converted to its decimal value. The decimal values of K were considered as 0, 1, 2, ..., 2^{K-1}. Thus, a good particle indicated a good substitution matrix. When the length of the secret message increased, the computational cost of calculating the peak signal to noise ratio (PSNR) value was also significantly increased. This was the major limitation of OLSBS [75].

Li and Wang [47] further addressed these shortcomings to achieve a higher security level for the secret message. An improved stego image quality was also attained as compared to JQTM and OLSBS. However, search for an optimal substitution matrix from within the number of substitution matrices still remained a big challenge. The combination of different optimization algorithms was also employed with image steganography [1, 21, 65]. However, the main focus of these methods was to improve the embedding capacity of secret text with reasonable image quality.

Thus, several optimization algorithms such as GA, PSO, ACO [21, 33, 53], etc. have been applied in the field of steganography to enhance the quality of stego images. These algorithms were computationally expensive for hiding large capacity of secret data and also very time consuming for locating the optimal solution. These limitations inspired us to work in this direction and explore CI and MRSLS optimization algorithms in the steganography field.

3.4 Summary

Most of the well-known metaheuristic optimization techniques such as SIO, ACO, PSO, GA, BA, FA, etc. are reviewed in this chapter. Furthermore, existing optimization algorithms used for steganography are also discussed. The focus is given to one of the emerging branches referred to as SIO. The SIO techniques are well received by the researchers and applied to many complex real-world problems for locating solutions. The upcoming popular optimization algorithms fall under this branch such as SCO, ICA, LCA, SEO, ECO, SLO, SGO, IA, SELO, CI, etc. are also discussed. An emerging and well-recognized socio inspired optimization algorithm referred to as CI, is reviewed. The literature on CI is also discussed to enhance the understanding of this algorithm. As per the current state of art discussed in the chapter, it is observed that the CI algorithm successfully locates optimized solutions of various types of engineering problems. However, the algorithm is not yet explored in the image steganography field. By considering its effectiveness, this algorithm is employed for this research to solve the practical concern of JPEG image steganography.

References

1. Aote, S.S., Raghuwanshi, M.M., Malik, L.G.: Improved particle swarm optimization based on natural flocking behavior. Arab. J. Sci. Eng. **41**(3), 1067–1076 (2016)
2. Aspnes, J., Chang, K., Yampolskiy, A.: Inoculation strategies for victims of viruses and the sum-of-squares partition problem. J. Comput. Syst. Sci. **72**(6), 1077–1093 (2006)
3. Azab, A., Naderi, B.: Greedy heuristics for distributed job shop problems. Procedia CIRP **20**, 7–12 (2014)

4. Azad, S.K., Kulkarni, A.J.: Structural optimization using a mutation-based genetic algorithm. Int. J. Optim. Civ. Eng. **2**(1), 81–101 (2012)
5. Badeau, P., Guertin, F., Gendreau, M., Potvin, J., Taillard, E.: A parallel tabu search heuristic for the vehicle routing problem with time windows. Transp. Res. Part C Emerg. Technol. **5**(2), 109–122 (1997)
6. Berend, D., Sapir, A., Solomon, S.: The tower of hanoi problem on path graphs. Discrete Appl. Math. **160**(10–11), 1465–1483 (2012)
7. Burke, E.K., Bykov, Y.: The late acceptance hill-climbing heuristic. Eur. J. Oper. Res. **258**(1), 70–78 (2017)
8. Cui, Z., Gao, X.: Theory and applications of swarm intelligence. Neural Comput. Appl. **21**(2), 205–206 (2012)
9. Dhavle, S.V., Kulkarni, A.J., Shastri, A., Kale, I.R.: Design and economic optimization of shell-and-tube heat exchanger using cohort intelligence algorithm. Neural Comput. Appl. (Springer) **30**(1), 111–125 (2018)
10. Dorigo, M., Birattari, M., Stitzle, T.: Ant colony optimization: artificial ants as a computational intelligence technique. IEEE Comput. Intell. Mag. **1**(4), 28–39 (2006)
11. Edward, J.S., Palaniappan, R., Ramakrishnan, S.: Imperceptibility—robustness tradeoff studies for ECG steganography using continuous ant colony optimization. Expert Syst. Appl. **49**, 123–135 (2016)
12. Eglese, R.W.: Simulated annealing: a tool for operational research. Eur. J. Oper. Res. **46**(3), 271–281 (1990)
13. Elbeltagi, E., Hegazy, T., Grierson, D.: Comparison among five evolutionary-based optimization algorithms. Adv. Eng. Inform. **19**(1), 43–53 (2005)
14. Fernandes, S., Lourenço, H.R.: A GRASP and branch-and-bound metaheuristic for the job-shop scheduling. In: European Conference on Evolutionary Computation in Combinatorial Optimization, Lecture Notes in Computer Science, (LNCS), vol. 4446, pp. 60–71. Springer (2007)
15. Foulds, L.R.: The heuristic problem-solving approach. J. Oper. Res. Soc. **34**(10), 927–934 (1983)
16. Gargari, E.A., Lucas, C.: Imperialist Competitive Algorithm: An Algorithm For Optimization Inspired By Imperialistic Competition. In: Evolutionary Computation, CEC, 2007 IEEE Congress, pp. 4661–4667. IEEE, Singapore (2007)
17. Ghasemi, E., Shanbehzadeh, J., Fassihi, N.: High capacity image steganography based on genetic algorithm and wavelet transform. In: Ao, S., Castillo, O., Huang, X. (eds) Intelligent Control and Innovative Computing. Lecture Notes in Electrical Engineering, vol 110, pp. 395–404. Springer, Boston (2012)
18. Ghosh, S., Mahanti, A.: Bidirectional heuristic search with limited resources. Inf. Process. Lett. **40**(6), 335–340 (1991)
19. Goldenberg, M.: The heuristic search research framework. Knowl.-Based Syst. **129**, 1–3 (2017)
20. Greco, R., Vanzi, I.: New few parameters differential evolution algorithm with application to structural identification. J. Traffic Transp. Eng. (English Edition) **6**(1), 1–14 (2019)
21. Hemanth, D.J., Umamaheswari, S., Popescu, D.E., Naaji, A.: Application of genetic algorithm and particle swarm optimization techniques for improved image steganography systems. Open Phys. **14**(1), 452–462 (2016)
22. Hosseini, S., Al Khaled, A.: A survey on the imperialist competitive algorithm metaheuristic: implementation in engineering domain and directions for future research. Appl. Soft Comput. **24**, 1078–1094 (2014)
23. Huan, T.T., Kulkarni, A.J., Kanesan, J.: Ideology algorithm: a socio-inspired optimization methodology. Neural Comput. Appl. **28**(1), 845–876 (2016)
24. Huang, J., Shi, Y.Q., Shi, Y.: Embedding image watermarks in DC components. IEEE Trans. Circuits Syst. Video Technol. **10**(6), 974–979 (2000)
25. Ikeda, S., Ooka, R.: Metaheuristic optimization methods for a comprehensive operating schedule of battery, thermal energy storage, and heat source in a building energy system. Appl. Energy **151**, 192–205 (2015)

26. Iyer, V.H., Mahesh, S., Malpani, R., Sapre, M.S., Kulkarni, A.J.: Adaptive range genetic algorithm: a hybrid optimization approach and its application in the design and economic optimization of shell-and-tube heat exchanger. Eng. Appl. Artif. Intell. **85**, 441–461 (2019)
27. Javid, A.A.: Anarchic society optimization: a human-inspired method. In: Evolutionary Computation, CEC, 2011 IEEE Congress, pp. 2586–2592. IEEE, New Orleans, USA
28. Kale, I.R., Kulkarni, A.J.: Cohort intelligence algorithm for discrete and mixed variable engineering problems. Int. J. Parallel Emergent Distrib. Syst. **33**(6), 627–662 (2017)
29. Kale, I., Kulkarni, A.J.: A socio-based cohort intelligence algorithm for engineering problems. In: Kulkarni, A.J., Singh, P.K., Satapathy, S.C., Husseinzadeh, K.A., Tai, K. (eds.), Socio-cultural inspired metaheuristics, Studies in Computational Intelligence, vol. 828, pp. 121–135. Springer (2019)
30. Kashan, A.H.: League championship algorithm: a new algorithm for numerical function optimization. In: Proceedings of international conference on soft computing and pattern recognition, pp. 43–48, Malacca, Malaysia, 4–7 Dec 2009
31. Kearfott, R.B.: An interval branch and bound algorithm for bound constrained optimization problems. J. Global Optim. **2**(3), 259–280 (1992)
32. Kennedy, J., Eberhart, R.: Particle swarm optimization. In: Proceedings of IEEE International Conference on Neural Networks, pp 1942–1948, University of Western Australia, Perth, Western Australia, 27 Nov–1 Dec 1995
33. Khamrui, A., Mandal, J.K.: A genetic algorithm based steganography using discrete cosine transformation (GASDCT). In: Proceedings of International Conference on Computational Intelligence: Modeling Techniques and Applications (CIMTA), pp. 105–111, University of Kalyani, West Bengal, India, 27–28 Sept 2013
34. Krishnasamy, G., Kulkarni, A.J., Paramesran, R.: A hybrid approach for data clustering based on modified cohort intelligence and K-means. Expert Syst. Appl. **41**(3), 6009–6016 (2014)
35. Kulkarni, A.J., Baki, M.F., Chaouch, B.A.: Application of the cohort intelligence optimization method to three selected combinatorial optimization problems. Eur. J. Oper. Res. **250**(2), 427–447 (2016)
36. Kulkarni, A.J., Kale, I.R., Tai, K.: Probability collectives for solving discrete and mixed variable problems. Int. J. Comput. Aided Eng. Technol. **8**(4), 325–361 (2016)
37. Kulkarni, A.J., Shabir, H.: Solving 0-1 knapsack problem using cohort intelligence algorithm. Int. J. Mach. Learn. Cybernet. **7**(3), 427–441 (2016)
38. Kulkarni, A.J., Krishnasamy, G., Abraham, A.: Cohort intelligence: a socio-inspired optimization method. In: Intelligent Systems Reference Library, vol. 114. Springer, ISBN: 978-3-319-44254-9 (2017)
39. Kulkarni, A.J., Durugkar, I.P., Kumar, M.: Cohort intelligence: a self supervised learning behavior. In: Proceedings of the 2013 IEEE International Conference on Systems, Man and Cybernetics, pp 1396–1400, IEEE Computer Society, Washington, DC, USA, 13–16 Oct 2013
40. Kulkarni, O., Kulkarni, N., Kulkarni, A., Kakandikar, G.: Constrained cohort intelligence using static and dynamic penalty function approach for mechanical components design. Int. J. Parallel Emergent Distrib. Syst. **33**(6), 570–588 (2016)
41. Kumar, M., Kulkarni, A.J., Satapathy, S.C.: Socio evolution & learning optimization algorithm: a socio-inspired optimization methodology. Future Gener. Comput. Syst. **81**, 252–272 (2018)
42. Kumar, M., Kulkarni, A.J.: Socio-inspired optimization metaheuristics: a review. In: Kulkarni, A.J., Singh, P.K., Satapathy, S.C., Husseinzadeh, K.A., Tai, K. (eds.) Socio-cultural Inspired Metaheuristics, Studies in Computational Intelligence, vol 828, pp. 241–265. Springer (2019)
43. Kuo, H.C., Lin, C.H.: Cultural evolution algorithm for global optimizations and its applications. J. Appl. Res. Technol. **11**(4), 510–522 (2013)
44. Langley, P.: Learning to search: from weak methods to domain-specific heuristics. Cognitive Sci. **9**(2), 217–260 (1985)
45. Li, L.J., Huang, Z.B., Liu, F.: A heuristic particle swarm optimization method for truss structures with discrete variables. Comput. Struct. **87**(7–8), 435–443 (2009)
46. Li, J., Sun, Q., Zhou, M., Yu, X., Dai, X.: Colored traveling salesman problem and solution. IFAC Proc. Volumes **47**(3), 9575–9580 (2014)

47. Li, X., Wang, J.: A steganographic method based upon jpeg and particle swarm optimization algorithm. Inf. Sci. **177**, 3099–3109 (2007)
48. Liu, Z.Z., Chu, D.H., Song, C., Xue, X., Lu, B.Y.: Social learning optimization (SLO) algorithm paradigm and its application in QoS-aware cloud service composition. Inf. Sci. **326**, 315–333 (2016)
49. Lv, W., He, C., Li, D., Cheng, S., Luo, S., Zhang, X.: Election campaign optimization algorithm. Procedia Comput. Sci. **1**(1), 1377–1386 (2010)
50. Marde, K., Kulkarni, A.J.: Optimum design of four mechanical elements using cohort intelligence algorithm. In: Kulkarni, A.J., Singh, P.K., Satapathy, S.C., Husseinzadeh, K.A., Tai, K. (eds.) Socio-cultural Inspired Metaheuristics, Studies in Computational Intelligence, vol. 828, pp 1–25. Springer (2019)
51. Moosavian, N.: Soccer league competition algorithm for solving knapsack problems. Swarm Evol. Comput. **20**, 14–22 (2015)
52. Moosavian, N., Roodsari, B.K.: Soccer league competition algorithm: a novel meta-heuristic algorithm for optimal design of water distribution networks. Swarm Evol. Comput. **17**, 14–24 (2014)
53. Nickfarjam, A.M., Azimifar, Z.: Image steganography based on pixel ranking and particle swarm optimization. In: Proceedings of the 16th CSI International Symposium on Artificial Intelligence and Signal Processing (AISP 2012). IEEE, Shiraz, Fars, Iran, 2–3 May 2012
54. Orlin, J.B., Punnen, A.P., Schulz, A.S.: Integer programming: optimization and evaluation are equivalent. In: Dehne, F., Gavrilova, M., Sack, J.R., Tóth, C.D. (eds.) Algorithms, and Data Structures, WADS 2009, Lecture Notes in Computer Science, vol. 5664, pp. 519–529. Springer, Berlin, Heidelberg (2009)
55. Passos, C.A.S.: A beam search based algorithm to solve flowshop scheduling problems with constraints on shared resources. In: IFAC Management and Control of Production and Logistics, pp. 675–679, Grenoble, France (2000)
56. Patankar, N.S., Kulkarni, A.J.: Variations of cohort intelligence. Soft. Comput. **22**(6), 1731–1747 (2018)
57. Qu, H., Yi, Z.: A new algorithm for finding the shortest paths using PCNNs. Chaos, Solitons Fractals **33**(4), 1220–1229 (2007)
58. Ray, T., Liew, K.M.: Society and civilization: an optimization algorithm based on the simulation of social behavior. IEEE Trans. Evol. Comput. **7**(4), 386–396 (2003)
59. Roy, R.V., Patel, V.: An improved teaching-learning-based optimization algorithm for solving unconstrained optimization problems. Scientia Iranica **20**(3), 710–720 (2013)
60. Roy, R., Laha, S.: Optimization of stego image retaining secret information using genetic algorithm with 8-connected PSNR. In: Proceedings of 19th International Conference on Knowledge Based and Intelligent Information and Engineering Systems, pp. 468–477, Singapore, 7–9 Sept 2015
61. Roychowdhury, P., Mehra, S., Devarakonda, R., Shrivastava, P., Basu, S., Kulkarni, A.J.: A self-organizing multi-agent cooperative robotic system: an application of cohort intelligence algorithm. In: Kulkarni, A.J., Singh, P.K., Satapathy, S.C., Husseinzadeh, K.A., Tai, K. (eds.) Socio-cultural Inspired Metaheuristics, Studies in Computational Intelligence, vol. 828, pp. 27–40. Springer (2019)
62. Sapre, M.S. Kulkarni, A.J., Chettiar, L., Deshpande, I., Piprikar, B.: Mesh Smoothing of Complex Geometry using Variations of Cohort Intelligence Algorithm (In press: Evolutionary Intelligence) (2018)
63. Sarmah, D., Kulkarni, A.J.: JPEG based steganography methods using cohort intelligence with cognitive computing and modified multi random start local search optimization algorithms. Inf. Sci. **430–431**, 378–396 (2018)
64. Sarmah, D., Kulkarni, A.J.: Image steganography capacity improvement using cohort intelligence and modified multi random start local search methods. Arab. J. Sci. Eng. **43**(8), 3927–3950 (2018)
65. Sarmah, D.K., Kulkarni, A.J.: Improved cohort intelligence-a high capacity, swift and secure approach on jpeg image steganography. J. Inf. Secur. Appl. **45**, 90–106 (2019)

66. Satapathy, S., Naik, N.: Social group optimization (SGO): a new population evolutionary optimization technique. Complex Intell. Syst. **2**(3), 173–203 (2016)
67. Selvakumar, A.I., Thanushkodi, K.: Optimization using civilized swarm: solution to economic dispatch with multiple minima. Electr. Power Syst. Res. **79**(1), 8–16 (2009)
68. Sen, A.K., Bagchi, A., Zhang, W.: Average-case analysis of best-first search in two representative directed acyclic graphs. Artif. Intell. **155**(1–2), 183–206 (2004)
69. Shah, P., Agashe, S., Kulkarni, A.J.: Design of fractional PID Controller using cohort intelligence method. Front. Inf. Technol. Electron. Eng. **19**(3), 437–445 (2018)
70. Shastri, A.S., Kulkarni, A.J.: Multi-cohort intelligence algorithm: an intra- and inter-group learning behavior based socio-inspired optimization methodology. Int. J. Parallel Emergent Distrib. Syst. **33**(6), 675–715 (2018)
71. Sivaraj, H., Gopalakrishnan, G.: Random walk based heuristic algorithms for distributed memory model checking. Electron. Notes Theor. Comput. Sci. **89**(1), 1–17 (2003)
72. Teo, T.H., Kulkarni, A.J., Kanesan, J., Chuah, J.H., Abraham, A.: Ideology algorithm: a socio-inspired optimization methodology. Neural Comput. Appl. **28**(1), 845–876 (2017)
73. Thompson, G.L.: An integral simplex algorithm for solving combinatorial optimization problems. Comput. Optim. Appl. **22**(3), 351–367 (2002)
74. Tuparov, G., Tuparova, D., Jordanov, V.: Teaching sorting and searching algorithms through simulation-based learning objects in an introductory programming course. In: 5th World Conference on Educational Sciences—WCES 2013, Procedia—Social and Behavioral Sciences, vol. 116, pp. 2962–2966 (2014)
75. Wang, R.Z., Lin, C.F., Lin, J.C.: Image hiding by optimal LSB substitution and genetic algorithm. Pattern Recogn. **34**(3), 671–683 (2001)
76. Wu, S.J., Chow, P.T.: Steady-state genetic algorithms for discrete optimization of trusses. Comput. Struct. **56**(6), 979–991 (1995)
77. Wu, S.J., Chow, P.E.: Genetic algorithms for nonlinear mixed discrete-integer optimization problems via metagenetic parameter optimizations. Eng. Optim. **24**(2), 137–159 (2007)
78. Xu, Y., Cui, Z., Zeng, J.: Social emotional optimization algorithm for nonlinear constrained optimization problems. In: Swarm, Evolutionary, and Memetic Computing, SEMCCO 2010, Lecture Notes in Computer Science, vol. 6466, pp. 583–890. Springer, Berlin, Heidelberg (2010)
79. Yang, X., He, X.: Why the firefly algorithm works? In: Yang, X.S. (eds) Nature-Inspired Algorithms and Applied Optimization. Studies in Computational Intelligence, vol. 744, pp. 245–259. Springer, Cham (2017)
80. Yang, X.: Bat algorithm and cuckoo search: a tutorial. In: Yang, X.S. (eds) Artificial Intelligence, Evolutionary Computing, and Metaheuristics. Studies in Computational Intelligence, vol. 427, pp. 421–434. Springer, Berlin, Heidelberg (2013)
81. Yuce, B., Packianather, M., Mastrocinque, E., Pham, D.T., Lambiase, A.: Honey bees inspired optimization method: the bees algorithm. Insects **4**(4), 646–662 (2013)

Chapter 4
Cohort Intelligence with Cognitive Computing (CICC) and Modified-Multi Random Start Local Search (M-MRSLS) Optimization Algorithms for JPEG Image Steganography (Approach 1 and Approach 2)

The JPEG image format [7] is commonly used in various steganography techniques such as least significant bit insertion, masking, and filtering, transformations, etc. The algorithm of CI is inspired by the natural and social tendency of learning from one another. It has already been applied and tested for solving unconstrained, constrained and NP-hard combinatorial problems. The details of the CI algorithm are explained in Sect. 3.2.2. Cognitive Computing (CC) [8] is an emerging area based on the platform of artificial intelligence and has various applications in the field of machine learning [2]. Another interesting optimization technique Multi Random Start Local Search (MRSLS) has already been studied for NP-hard combinatorial problems. The performance of the algorithm in terms of quality of solution was found comparable to CI.

Considering two important aspects of image steganography i.e. picture quality and high data hiding capacity, a research effort was made to propose two novel image steganography techniques based on JPEG compression. The first technique combined CI with CC referred to as CICC and the second one modified MRSLS optimization algorithm referred to as M-MRSLS. The framework of CI was applied to JPEG image steganography. However, for JPEG based image steganography methods, as the capacity of secret text increases and/or the size of the image increases, the CI algorithm exhibited slow convergence and increased the computational cost. In order to diminish this limitation of CI, the CC concept was combined with CI referred to as CICC and applied to JPEG image compression for data hiding. CICC was used to search and select the optimal substitution matrix which was further used to transform the secret message into ciphertext.

MRSLS optimization method is also proposed by Kulkarni et al. [4] to solve the constrained combinatorial problems. The details of the original MRSLS methodology are presented in Sect. 3.2.3. The MRSLS is a technique based on the pairwise solution interchange approach. In order to search for better solutions, the elements of two adjacent solutions were randomly swapped. As the initial solution was randomly generated in MRSLS, sometimes these random solutions could be infeasible and

D. K. Sarmah et al., *Optimization Models in Steganography Using Metaheuristics*, Intelligent Systems Reference Library 187, https://doi.org/10.1007/978-3-030-42044-4_4

may require constructing the initial feasible solutions. Also, the final solution was generated by only considering the behavior of a pair of solutions which may lead the solution to stuck in local optima. Thus, a modified version of the MRSLS referred to as M-MRSLS was developed and applied to JPEG image compression for data hiding which was further used to select the optimal substitution matrix. In M-MRSLS, a random solution was generated by observing the behavior of other candidates. Thus, both the optimization algorithms referred to as CICC and M-MRSLS were used for experimentation and further achieved better results in the steganography domain. The detailed procedure including embedding and extraction of secret text for CICC and M-MRSLS algorithm is presented in Sect. 4.1. The validation of the algorithms was also accomplished by comparing different evaluation parameters with the existing algorithms such as JQTM, the method proposed by Li et al. [6], non-optimal substitution method. Section 4.2 presents the entire discussion on results. The concluding remarks of this chapter are presented in Sect. 4.3.

4.1   CICC/M-MRSLS with Image Steganography

The framework of CICC and M-MRSLS with JPEG image steganography is explained in this section. This section is divided into two sections. Embedding procedure of secret text is explained in Sect. 4.1.1 and the extraction procedure to extract the secret message is explained in Sect. 4.1.2.

4.1.1   Embedding Procedure with Example

The flowchart for the embedding procedure is given in Fig. 4.1. The procedure was divided into three phases. (i) In Phase 1, a secret message was accepted and transformed into ciphertext using different substitution matrices. (ii) In Phase 2 either of the CICC or M-MRSLS optimization technique was employed to identify the optimal substitution matrix and its corresponding transformed message. (iii) In Phase 3, the transformed secret message was hidden into low to middle-frequency components of the quantized DCT coefficients of each 8×8 block of the image matrix. The implementation of the complete embedding procedure is explained in detail.

(i) **Phase 1 (Transformation of secret text)**

In this phase, a secret message was accepted and converted into a smaller number of k bits. This phase was inspired by the OLSBS method [9]. k LSBs of the chosen DCT coefficients in each block were further used to embed the k secret bits. The range of the decimal values of k bits secret messages was from 0 to $2^k - 1$. Substitution matrix (M) was used to transform the original secret bits into ciphertext. The substitution matrix was represented as follows: $M = \{m_{ij}, 0 \leq i, j \leq 2^k - 1\}$

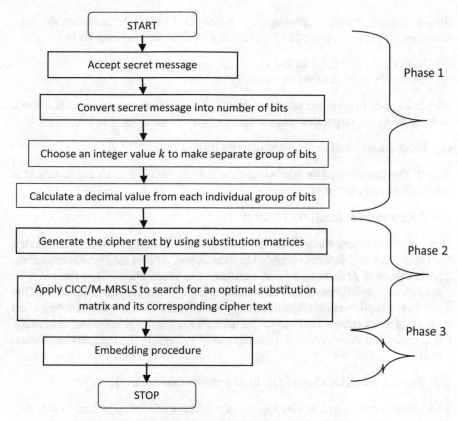

Fig. 4.1 Overall embedding procedure flowchart

$$\text{where } m_{ij} = \begin{cases} 1, & \text{if i replace j} \\ 0, & \text{otherwise} \end{cases} \qquad (4.1.1a)$$

These substitution matrices were either in the form of Identity matrix (I) and its variations. There was a total $2^k!$ substitution matrices, denoted as $M_1, \ldots, M_{2^k!}$. The value of k was considered as 2 for simplicity which gave the decimal values of 2 bits secret messages in the range of 0 to 3. Therefore, M was considered the size of 4×4 for this case. The total possible combinations of substitution matrices were 24.

An illustration of transforming a secret message is described below in the following steps:

Step 1: Assume substitution matrix M in the form of a variation of the identity matrix.

$$\text{(a)} \quad M: \begin{array}{c} 0 \\ 1 \\ 2 \\ 3 \end{array} \begin{bmatrix} 0 & 1 & 0 & 0 \\ 1 & 0 & 0 & 0 \\ 0 & 0 & 1 & 0 \\ 0 & 0 & 0 & 1 \end{bmatrix}$$

Step 2: Assume the secret message is of 8 bits {1 0 1 1 0 0 0 1}. The secret message is divided into four chunks of 2 bits and converts them into decimal form.

(a)   Secret message: {1 0 1 1 0 0 0 1}
(b)   Decimal values of secret message: {2310}

Step 3: Replace the decimal values of secret message which is considered as the row index of M by the respective column index where the value of M is found as 1.

(c)   Substitution result of (b) using (a): {2 3 0 1}

Step 4: The transformed decimal values are then converted into binary form and is considered as a result of M.

(d)   Binary value of (c): {1 0 1 1 0 0 0 1}

These new transformed bits of the message are then to be hidden into the selected DCT coefficients of the JPEG image. The substitution process used here is reversible. The transpose of M is to be used to transform the substitution. Since there can be a number of possibilities of M which direct to a stego image with different quality. Therefore, an optimization algorithm is required to search for the optimal substitution matrix which is explained in Phase 2 (Identification of Optimal Substitution Matrix). CICC optimization algorithm is used separately to identify the optimal substitution matrix.

(ii)   **Phase 2 (Identification of Optimal Substitution Matrix)**

This phase is considered to elaborate on the process of two optimization methods for selecting the optimal substitution matrix and is divided into two parts:

(a)   Part 1: Find the optimal substitution matrix using CICC
(b)   Part 2: Find the optimal substitution matrix using M-MRSLS

Both the parts i.e. Part 1 and Part 2 are explained through Steps, Flowchart, and a sample illustration. Flowchart for CICC and M-MRSLS are shown in Figs. 4.2 and 4.4, respectively. Further, their illustrations are shown under its consecutive sections i.e. Part 1.1 and Part 2.1, respectively.

Part 1 (Optimal substitution matrix using CICC)

The flow chart of CICC is shown in Fig. 4.2. In the context of CI algorithm (For details refer to Sect. 3.2.2), each substitution matrix (M) was considered as a candidate in the cohort. There were 24 candidates for selecting the value of $k = 2$ where randomly 4 candidates were chosen out of 24 for simplicity. In order to evaluate the quality of stego image, a fitness function i.e. PSNR was considered. The value of PSNR was inversely dependent upon Mean square Error (MSE) which calculates the difference between the pixel values of the stego image and the corresponding cover image. As described in the CI algorithm in Sect. 3.2.2, each candidate in the cohort has certain qualities that contribute to the overall behavior of the cohort. In the proposed

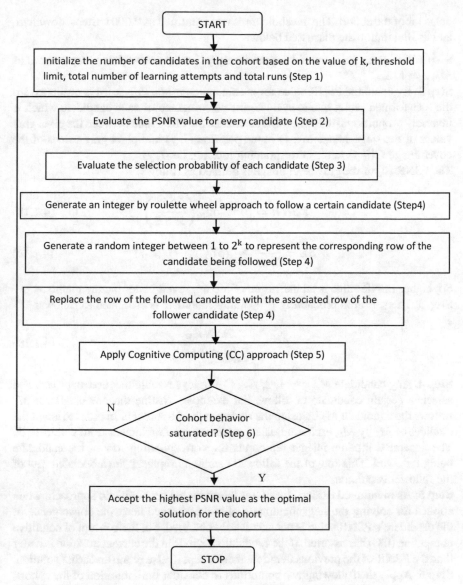

Fig. 4.2 Cohort intelligence with cognitive computing (CICC) flowchart (approach 1)

algorithm, the quality of each candidate was considered as the position of the row in M that has the value 1. There was a total of 4 qualities for each candidate, accordingly, matrices having a size of 4×4 were considered. A total of 20 runs were considered for implementation purposes. Each run was having a maximum of 40 learning attempts to check the total number of attempts taken by the candidates to improve the overall behavior of the cohort. The threshold limit was identified as 0.0001. Steps, flowchart, and its illustration are discussed below:

Step 1: Generate the C candidates randomly where any candidate is represented as $M_c, c = 1, 2, \ldots, C$.

Step 2: Evaluate the PSNR value for each candidate M_c. PSNR [6] is evaluated for the transformed secret bits to evaluate the performance of each candidate which is inversely proportional to MSE. MSE calculates the difference between the pixel gray values of the stego image $S(i, j)$ in the position (i, j) and pixel gray values of the cover image $C(i, j)$ in the corresponding location i.e.(i, j).

The PSNR [6] of the gray level image is defined as follows:

$$PSNR = 10 \times \log_{10}\left(\frac{255^2}{MSE}\right) \tag{4.1.1b}$$

$$MSE = \frac{1}{WH} \sum (S(i, j) - C(i, j))^2 \tag{4.1.1c}$$

So, higher PSNR value is an indication of the improved/better picture quality.

Step 3: The selection probability P^{M_c} of any candidate is evaluated as follows:

$$P^{M_c} = \frac{PSNR_{M_c}}{\sum_{c=1}^{C} PSNR_{M_c}} \tag{4.1.1d}$$

Step 4: Any candidate $M_c, c = 1, 2, \ldots, C$ employs a roulette wheel approach and selects a certain candidate to follow. For example, assume that the candidate M_1 follows the candidate M_2 by using the roulette wheel approach. In order to adapt the qualities of M_2 by M_1, each candidate generates a random integer from within $1-2^k$. The generated random integer represents the corresponding row of the candidate being followed. This row of the followed candidate replaces the associated row of the follower candidate.

Step 5: As mentioned earlier, since CI is based on the probabilistic approach, when applied for solving the steganography problem addressed here, the convergence of the candidate's PSNR value is not necessarily obtained. So, the concept of cognitive computing (CC) has resorted. If the candidate's PSNR in the current iteration is better than the PSNR of the previous iteration, then accept it, else retain an earlier solution.

Step 6: Apply the following two conditions to consider the saturation of the cohort;

(a) If the maximum number of learning attempts is reached.

(b) If any candidate's behavior in the cohort has no significant improvement i.e. the difference between the PSNR values of any candidate in the successive learning attempts does not change significantly.

All the conditions executed simultaneously to achieve the saturation point which helped us to provide the optimal solution. CICC flowchart as shown in Fig. 4.2 describes the complete process step by step. Further, the whole process of CICC is explained through a sample illustration in Part 1.1 as shown in Fig. 4.3 wherein a

Candidates

$$
a) \ M_1 \quad \begin{bmatrix} 0 & 1 & 0 & 0 \\ 1 & 0 & 0 & 0 \\ 0 & 0 & 1 & 0 \\ 0 & 0 & 0 & 1 \end{bmatrix} \quad b) M_2 \quad \begin{bmatrix} 1 & 0 & 0 & 0 \\ 0 & 0 & 1 & 0 \\ 0 & 1 & 0 & 0 \\ 0 & 0 & 0 & 1 \end{bmatrix} \quad c) \ M_3 \quad \begin{bmatrix} 0 & 0 & 0 & 1 \\ 0 & 1 & 0 & 0 \\ 1 & 0 & 0 & 0 \\ 0 & 0 & 1 & 0 \end{bmatrix} \quad d) \ M_4 \quad \begin{bmatrix} 0 & 0 & 1 & 0 \\ 0 & 0 & 0 & 1 \\ 1 & 0 & 0 & 0 \\ 0 & 1 & 0 & 0 \end{bmatrix}
$$

a) M_1 b)M_2 c) M_3 d) M_4

Evaluate the *PSNR* value of each candidate $PSNR_1 = 38.1837$, $PSNR_2 = 38.2091$, $PSNR_3 = 38.2417$, and $PSNR_4 = 38.101$

Calculate the total value of *PSNR* $PSNR_T = 152.7355$

Evaluate the selection probability of each candidate. $P_1 = \dfrac{PSNR_1}{PSNR_T}$, $P_2 = \dfrac{PSNR_2}{PSNR_T}$, $P_3 = \dfrac{PSNR_3}{PSNR_T}$, $P_4 = \dfrac{PSNR_4}{PSNR_T}$

Thus,

$$P_1 = \frac{38.1837}{152.7355} = 0.2499,$$

$$P_2 = \frac{38.2091}{152.7355} = 0.2501, P_3 = \frac{38.2417}{152.7355}$$

$$= 0.2503, \qquad P_4 = \frac{38.101}{152.7355} = 0.2494$$

Calculate the candidate's cumulative probability. $P_{1_{cu}} = 0.2499, P_{2_{cu}} = 0.5, P_{3_{cu}} = 0.7503, P_{4_{cu}} = 0.9997$

Every candidate generates 4 values between 0 and 1 using roulette wheel approach to follow the certain candidate. $\{0.3333, 0.1123, 0.8234, 0.3445\}$

Implies

M_1 candidate follows M_2 candidate,

M_2 candidate follows M_1 candidate,

M_3 candidate follows M_4 candidate and

M_4 candidate follows M_2 candidate.

Every candidate generates a random integer between 1 and 4 which corresponds to the row number of the followed candidate. $\{2, 3, 1, 4\}$

Fig. 4.3 Illustration a grayscale image using CICC

grayscale image of Lena having size 256×256 is considered.

Replace the row_2 of the candidate M_1 with row_2 of the candidate M_2, row_3 of the candidate M_2 with row_3 of the candidate M_1, row_1 of the candidate M_3 with row_1 of the candidate M_4 and row_4 of the candidate M_4 with row_4 of the candidate M_2 in order to adapt the quality of the followed candidate by the follower candidate

Candidate M_1: $\begin{bmatrix} 0 & 1 & 0 & 0 \\ 1 & 0 & 0 & 0 \\ 0 & 0 & 1 & 0 \\ 0 & 0 & 0 & 1 \end{bmatrix}$ row_2 of the candidate M_1: $\{1\ 0\ 0\ 0\}$

Candidate M_2: $\begin{bmatrix} 1 & 0 & 0 & 0 \\ 0 & 0 & 1 & 0 \\ 0 & 1 & 0 & 0 \\ 0 & 0 & 0 & 1 \end{bmatrix}$ row_2 of the candidate M_2: $\{0\ 0\ 1\ 0\}$

After replacement, the new candidate $M_1{}'$: $\begin{bmatrix} 0 & 1 & 0 & 0 \\ 0 & 0 & 1 & 0 \\ 0 & 1 & 0 & 0 \\ 0 & 0 & 0 & 1 \end{bmatrix}$

The new 4 candidates in the cohort

$\begin{bmatrix} 0 & 1 & 0 & 0 \\ 0 & 0 & 1 & 0 \\ 0 & 1 & 0 & 0 \\ 0 & 0 & 0 & 1 \end{bmatrix}\begin{bmatrix} 1 & 0 & 0 & 0 \\ 1 & 0 & 0 & 0 \\ 0 & 0 & 1 & 0 \\ 0 & 1 & 0 & 0 \end{bmatrix}\begin{bmatrix} 0 & 0 & 1 & 0 \\ 0 & 1 & 0 & 0 \\ 1 & 0 & 0 & 0 \\ 0 & 0 & 1 & 0 \end{bmatrix}\begin{bmatrix} 0 & 0 & 0 & 1 \\ 0 & 0 & 0 & 1 \\ 1 & 0 & 0 & 0 \\ 0 & 0 & 0 & 1 \end{bmatrix}$

a) $M_1{}'$ (b)$M_2{}'$ (c)$M_3{}'$ (d) $M_4{}'$

Calculate the PSNR values of the 4 new candidates

$PSNR'_1 = 39.1833, PSNR'_2 = 38.1091, PSNR'_3 = 38.2017,$ and $PSNR'_4 = 38.2037$

Apply CC approach in all the PSNR values of every candidate and compare these values from the current to the previous iteration.

$M_1{}' > M_1$
$M_2{}' < M_2$
$M_3{}' < M_3$
$M_4{}' > M_4$

Accept the higher PSNR values for every candidate indicates the PSNR value of the current iteration.

$39.1833, 38.2091, 38.2417, 38.2037$

Continue the process until the saturation condition is reached as mentioned in Step 6 as depicted in Figure 2.

Fig. 4.3   (continued)

Part 1.1 (A sample Illustration of a grayscale image using CICC)

Let us consider the total number of candidates $M_c, c = 1, 2, \ldots, 2^k$ as $k = 2$. Thus, the total number of candidates in the cohort is 4. A grayscale image of Lena having size 256×256 is considered for this illustration wherein a single iteration is shown in Fig. 4.3.

Part-2 (Optimal substitution matrix using M-MRSLS)

The mathematical formulation for M-MRSLS methodology is explained below in detail. An algorithm flowchart and an illustration are shown in Figs. 4.4 and 4.5, respectively. Please refer to Sect. 3.2.3 for the description of the original MRSLS algorithm [4]. In the context of MRSLS, each substitution matrix (M) was considered as a solution in a set $\{M_p, p = \{1, 2, 3, \ldots, P\}$ where $P = 2^k!\}$. There was a total of 24 solutions for selecting the value of $k = 2$. A total of 4 solutions were selected randomly out of 24 for simplicity. In order to evaluate the quality of a stego image, a fitness function PSNR (refer to Phase 2 Part 1 for more detail) was considered. The quality of a stego image was dependent upon the behavior of a set M_p. As described the MRSLS algorithm in Sect. 3.2.3, each solution in the set M_p generates

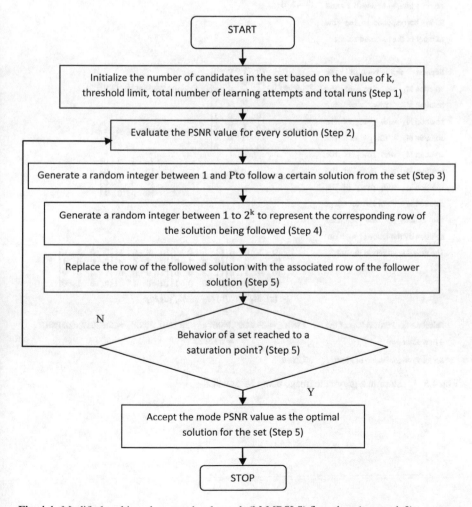

Fig. 4.4 Modified multi random start local search (M-MRSLS) flow chart (approach 2)

Solutions

$$\begin{bmatrix} 0 & 1 & 0 & 0 \\ 1 & 0 & 0 & 0 \\ 0 & 0 & 1 & 0 \\ 0 & 0 & 0 & 1 \end{bmatrix} \begin{bmatrix} 1 & 0 & 0 & 0 \\ 0 & 0 & 1 & 0 \\ 0 & 1 & 0 & 0 \\ 0 & 0 & 0 & 1 \end{bmatrix} \begin{bmatrix} 0 & 0 & 0 & 1 \\ 0 & 1 & 0 & 0 \\ 1 & 0 & 0 & 0 \\ 0 & 0 & 1 & 0 \end{bmatrix} \begin{bmatrix} 0 & 0 & 1 & 0 \\ 0 & 0 & 0 & 1 \\ 1 & 0 & 0 & 0 \\ 0 & 1 & 0 & 0 \end{bmatrix}$$

a) M_1 b) M_2 c) M_3 d) M_4

Evaluate the PSNR value of each solution

$PSNR_1 = 38.2837$, $PSNR_2 = 38.2091$, $PSNR_3 = 38.2175$, and $PSNR_4$
$= 38.101$

Select a random value of PSNR   38.2175

Set generates a single random value between 1 and 4 to follow a certain solution.

3

Implies

$\{M_1, M_2, M_3, M_4\}$ follows a solution M_3

Every solution generates a random integer between 1 and4 which corresponds to the row number of the followed solution.

$\{2, 3, 1, 4\}$

Replace the row_2 of the solution M_1 with row_2 of the solution M_3, row_3 of the solution M_2 with row_3 of the solution M_3, row_1 of the solution M_3 with row_1 of the solution M_3 and row_4 of the solution M_4 with row_4 of the solution M_3 in order to adapt the quality of the followed solution by the follower solution

Solution M_1: $\begin{bmatrix} 0 & 1 & 0 & 0 \\ 1 & 0 & 0 & 0 \\ 0 & 0 & 1 & 0 \\ 0 & 0 & 0 & 1 \end{bmatrix}$ row_2 of the solution M_1: $\{1\ 0\ 0\ 0\}$

Solution M_3: $\begin{bmatrix} 0 & 0 & 0 & 1 \\ 0 & 1 & 0 & 0 \\ 1 & 0 & 0 & 0 \\ 0 & 0 & 1 & 0 \end{bmatrix}$ row_2 of the solution M_3: $\{0\ 1\ 0\ 0\}$

After replacement, the new solution M_1': $\begin{bmatrix} 0 & 1 & 0 & 0 \\ 0 & 1 & 0 & 0 \\ 0 & 0 & 1 & 0 \\ 0 & 0 & 0 & 1 \end{bmatrix}$

The new 4 solutions in the set

$$\begin{bmatrix} 0 & 1 & 0 & 0 \\ 0 & 1 & 0 & 0 \\ 0 & 0 & 1 & 0 \\ 0 & 0 & 0 & 1 \end{bmatrix} \begin{bmatrix} 1 & 0 & 0 & 0 \\ 0 & 0 & 1 & 0 \\ 1 & 0 & 0 & 0 \\ 0 & 0 & 0 & 1 \end{bmatrix} \begin{bmatrix} 0 & 0 & 0 & 1 \\ 0 & 1 & 0 & 0 \\ 1 & 0 & 0 & 0 \\ 0 & 0 & 1 & 0 \end{bmatrix} \begin{bmatrix} 0 & 0 & 1 & 0 \\ 0 & 0 & 0 & 1 \\ 1 & 0 & 0 & 0 \\ 0 & 0 & 1 & 0 \end{bmatrix}$$

(a) M_1' (b) M_2' (c) M_3' (d) M_4'

Calculate the PSNR values of the 4 new solutions

$PSNR'_1 = 39.2169$, $PSNR'_2 = 38.1091$, $PSNR'_3 = 38.2017$, and $PSNR'_4$
$= 39.3037$

Select a random value of PSNR   39.2169

Fig. 4.5 Illustration a grayscale image using M-MRSLS

Continue the process until the saturation condition is reached as mentioned in Step 6 of Part 2.

Selected random PSNR values	38.2175, 38.2169, 38.2235, 38.2207, 38.2207, 38.2260, 38.2221,
from each iteration	38.2221, 38.2176, 38.2176, 38.2163, 38.2163, 38.2154, 38.2154,
	38.2154, 38.2154, 38.2173, 38.2154, 38.2164, 38.2091, 38.2148,
	38.2154, 38.2158, 38.2164, 38.2229, 38.2270, 38.2197, 38.2232,
	38.2348, 38.2232, 38.2348, 38.2229, 38.2221, 38.2229, 38.2275

| Accept the mode PSNR value | 38.2154 |

Fig. 4.5   (continued)

a neighboring solution by using a pair-wise interchange approach. However, M-MRSLS generates a solution on a random basis, depending upon the behavior of every associated solution. The quality of each solution makes the overall behavior of the set and it is considered as the position of the row in M that has the value 1. Each solution is in the form of the matrices having size 4×4, thus there are a total of 4 numbers of qualities for each solution. A total of 20 runs were considered for implementation purposes. Each run was having a maximum of 40 iterations to check the total number of attempts taken to improve the overall behavior of the set:

Step 1: Generate the P solutions randomly where any solution is represented as $\{M_p, p = \{1, 2, 3, \ldots, P) \ where \ P = 2^k!\}$.

Step 2: Evaluate the PSNR [6] value for each solution M_p. PSNR is evaluated for the transformed secret bits to evaluate the performance of each solution in the set which is inversely proportional to MSE. PSNR and MSE are shown in Eqs. 4.1.1b and 4.1.1c, respectively.

Step 3: Every solution $M_p, p = 1, 2, \ldots, P$ generates a random integer between 1 and p and follows a solution associated with the generated integer. For example, assume that each solution $M_p, p = 1, 2, \ldots, P$ generates a random integer 3 which implies that each solution of the set follows the solution M_3.

Step 4: In order to adapt the qualities of M_3, every solution generates a random integer from within $1-2^k$. The generated random integer represents the corresponding row of the solution being followed. This row of the followed solution replaces the associated row of the follower solution.

Step 5: Since large randomization is introduced in the M-MRSLS approach, the concept of statistical Mode is included in the algorithm to select the optimized fitness value amongst the generated feasible solutions. The Mode is the number that occurs most frequently within a set of numbers. Mode helps identify the most common or frequent occurrence of a characteristic. In order to achieve the optimized fitness value, the total number of attempts i.e. 40 are considered as the saturation point. The same process is repeated until Step 5 until the saturation point is reached.

Flowchart as shown in Fig. 4.4 brings more clarity for the M-MRSLS algorithm. Further, the same algorithm is explained using illustration in Fig. 4.5 wherein sample values from the implementation are considered for a single iteration. Once the optimization matrix is achieved either by CICC or M-MRSLS, a transformed secret message is hidden into the frequency components of DCT coefficients of every block of a grayscale image. The hiding procedure of the transformed secret message is shown in the next section i.e. Phase 3 (Embedding procedure of transformed secret message) after the illustration of M-MRSLS.

Part 2.1 (A sample Illustration of a grayscale image using M-MRSLS)

Let us consider the total number of solutions in a set M_p, $p = 1, 2, \ldots, 2^k$ as $k = 2$. Thus, the total number of solutions in the set is 4. A grayscale image of Lena having size 256×256 is considered for this illustration wherein a single iteration is shown in Fig. 4.5.

(iii) Phase 3 (Embedding Procedure of Transformed Secret Message)

In this phase, the hiding procedure of optimal transformed secret message generated by the optimal substitution matrix (as discussed in Phase 2) is explained. Low to middle-frequency components of the quantized DCT coefficients of each block were selected for hiding. As described in Sect. 2.2, an image is fragmented by DCT into three frequency bands namely the high, middle and low-frequency bands. A blocking artifact [3] would easily be created if the DC coefficients are modified. Blocking artifacts introduce the distortion of the visual quality of the image and is considered a substantial problem in DCT based image compression [11]. Therefore, AC coefficients were considered for hiding purposes, especially low to middle-frequency coefficients. If the high-frequency coefficients are selected for hiding, it may expose the secret information through compression and noise attacks as high-frequency components are easily targeted by the attacker [5]. Thus, the overall embedding procedure is dependent upon hiding transformed secret message which was identified either by CICC/M-MRSLS methodology. The diagram of the embedding procedure using CICC/M-MRSLS is shown in Fig. 4.6. The hiding of the secret text was done after the quantization step as shown in Fig. 2.3a. The modified quantization table [6] was selected as shown in Table 6.2. The entire JPEG image compression procedure is explained in Sect. 2.2.

4.1.2 Extraction Procedure

The extraction procedure of secret text was implemented at the receiver side. The receiver accepted the stego image and the optimal substitution matrix. The receiver's aim is to retrieve the secret text from the stego image. Based on the optimization method i.e. either CICC or M-MRSLS, the transpose operation was applied to the optimal substitution matrix to get the secret message. The whole extraction procedure is described below in detail and shown in Fig. 4.7.

There are six steps involved in the extraction procedure as shown in Fig. 4.7.

Step 1: Apply entropy decoding (refer to Sect. 2.2 for details) to decode the stego image and to retrieve the 8×8 blocks of a stego image.

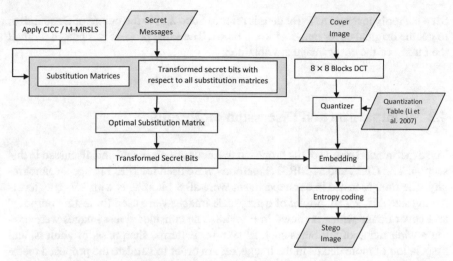

Fig. 4.6 Embedding procedure of CICC/M-MRSLS

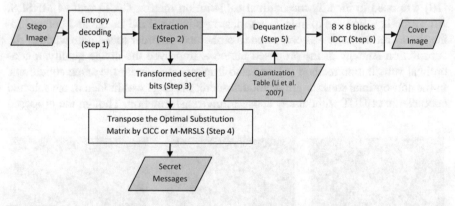

Fig. 4.7 Extraction procedure

Step 2: Extract the secret bits from 8×8 blocks of the stego image. Extracting order is considered as same as the embedding order (refer to Sect. 2.2 for details).

Step 3: Retrieve the transformed secret bits.

Step 4: Transpose the optimal substitution matrix determined either by CICC or M-MRSLS. Pass the transformed secret bits generated from Step 3 to the transpose of the optimal substitution matrix to evaluate the original secret bits.

Step 5: In order to retrieve the original cover image, after extracting secret bits from 8×8 blocks of the stego image, the output of the extraction procedure passes on to Dequantizer (For details refer to Sect. 2.2).

Step 6: Apply inverse DCT (for details refer to Sect. 2.2) to the output of Dequantizer to get the original cover image of 8×8 block. The same process is repeated until all the blocks of the cover image are obtained.

4.2   Comparison and Discussion on Results

The experimental results of the proposed methods are presented and discussed in this section. The CICC and M-MRSLS methods were used for JPEG image steganography. The images tested in our experiment were all 8-bit images with 256 gray levels having size 256×256. A total of 6 grayscale images were used for testing purposes as a cover image which is shown in Fig. 4.8. The considered test images were having a wide range of attributes such as texture, patterns, sharpness, brightness, and association of more details in the image, etc. In order to validate the proposed methods, a comparative analysis was carried out amongst different methods under the same circumstances. Methods considered were JQTM, JPEG_PSO [6] where PSO [10] was used to identify the optimal substitution matrix, CICC, and M-MRSLS. All methods were used to hide secret messages in the JPEG image and before hiding the secret message was converted to ciphertext. In order to validate whether the substitution strategy of the proposed methods improved the image quality, a non-optimal substitution method [6] was also implemented under the same conditions. In the non-optimal substitution method, a secret message was hidden in the selected coefficients of DCT without any substitution/transformation. Though the proposed

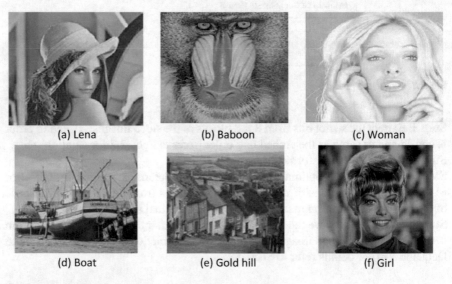

(a) Lena   (b) Baboon   (c) Woman

(d) Boat   (e) Gold hill   (f) Girl

Fig. 4.8   Six cover images with size 256×256

Table 4.1 Comparison: image quality (PSNR in decibel (dB))

Images	Method				
	JQTM	Non-optimal substitution	JPEG_PSO	CICC	M-MRSLS
Lena	35.8606	37.7840	37.0600	38.2417	38.2275
Baboon	29.5103	29.1646	31.2800	29.9358	29.9350
Woman	39.9383	42.1203	35.7100	42.3867	42.4212
Boat	36.2393	38.5884	36.2900	38.6128	38.5428
Goldhill	36.6300	41.0400	36.7800	41.6636	41.6362
Girl	29.3303	29.5103	38.0200	30.1102	30.1102

algorithms worked differently for different parameters on different test images, however, the better results achieved in comparison to the existing algorithms i.e. JQTM [1], Non-Optimal Substitution [6] and JPEG_PSO [6].

For every image, 20 independent CICC and M-MRSLS runs were conducted. Each run was having 40 learning attempts/iterations. Results were compared in terms of best, median and worst PSNR, function evaluations, standard deviation and elapsed time. Table 4.1 presents the comparison of PSNR between JQTM, Non-optimal substitution, Li et al. [6], CICC and M-MRSLS methods. The proposed methods achieved better quality in comparison to other methods for the test images. Also, result in analysis between the proposed methods exhibited better image quality of CICC in comparison to M-MRSLS for most of the test images. Best, Median and worst cases of image quality for the proposed methods are shown in Table 4.2. The best and worst image qualities for the proposed methods were almost comparable for the same test images. The standard deviation of PSNR for the proposed methods was also calculated and compared with JPEG_PSO [6] method as shown in Table 4.3. This validated the fact that the proposed methods were more robust than JPEG_PSO. Also, the standard deviation of PSNR for CICC was found better than M-MRSLS. It indicated more robustness towards CICC than M-MRSLS. Table 4.4 presents the comparison of function evaluations for the proposed methods. Though there were some variations found while judging these algorithms among themselves with respect to evaluation parameters, the improved results were observed in comparison to the other existing algorithms considered in our work. Computational time was also calculated for the proposed methods, JQTM, non-optimal substitution method and JPEG_PSO method as shown in Table 4.5. The computational time for best, median and worst case was also calculated for the proposed methods: JQTM, Non-optimal substitution method and JPEG_PSO method as shown in Table 4.6. Since there were 40 learning attempts/iterations used for the execution of the proposed algorithms to achieve more refine and optimal result, the computational time was increased for both the methods in comparison to the other methods i.e. JQTM, Non-optimal substitution method, and JPEG_PSO method. Though the same number of learning attempts/iterations and the same number of runs were used for both the proposed methods, CICC was found better than M-MRSLS for Baboon, Woman, Boat and Gold Hill Test images. A comparison for capacity (in bits) can be seen in Table 4.7.

Table 4.2 Comparison: best, median and worst image quality (PSNR in dB)

Images		Methods	
		CICC	M-MRSLS
Lena	Best	38.2417	38.2417
	Median	38.2417	38.2275
	Worst	38.1725	38.1725
Baboon	Best	29.9376	29.9376
	Median	29.9358	29.9350
	Worst	29.9340	29.9335
Woman	Best	42.4212	42.4212
	Median	42.3867	42.4212
	Worst	42.3867	42.3625
Boat	Best	38.6204	38.6204
	Median	38.6128	38.5428
	Worst	38.5592	38.5428
Goldhill	Best	41.6636	41.6636
	Median	41.6636	41.6362
	Worst	41.6225	41.6289
Girl	Best	30.1151	30.1151
	Median	30.1102	30.1102
	Worst	30.1097	30.1095

Table 4.3 Comparison: standard deviation of PSNR

Images	Methods		
	Standard deviation		
	JPEG_PSO	CICC	M-MRSLS
Lena	0.0300	0.01814	0.01968
Baboon	0.0100	0.0008	0.0013
Woman	0.0200	0.00796	0.0224
Boat	0.0200	0.0137	0.0281
Goldhill	0.0300	0.0138	0.0105
Girl	0.0300	0.0012	0.0019

JQTM method hides 2 bits of secret message in the selected DCT coefficients and there is a total of 26 coefficients in each block selected for hiding. Thus, each block can hide $26 \times 2 = 52$ secret bits. In Li et al. [6], a total of 36 coefficients were selected for hiding, therefore the secret bits hiding capacity for this method was $36 \times 2 = 72$ bits. The proposed methods used the same quantization table as used in Li et al. [6]. Therefore, the hiding capacities of the proposed methods were the same as Li et al. [6]. Stego Images with PSNR for CICC and M-MRSLS are presented in Figs. 4.9 and 4.10, respectively.

Table 4.4 Comparison: standard deviation of function evaluations

Images	Methods	
	Function evaluations	
	CICC	M-MRSLS
Lena	36	31
Baboon	29	32
Woman	33	28
Boat	29	32
Goldhill	27	31
Girl	35	28

Table 4.5 Comparison: standard deviation of computational time

Images	Methods				
	Computational time				
	CICC	M-MRSLS	JPEG_PSO	JQTM	Non-optimal substitution
Lena	99.8138	79.3653	35.5000	2.4448	2.6235
Baboon	101.2855	108.0255	35.7000	3.2807	3.2176
Woman	86.3609	101.6035	35.5000	2.4350	2.6852
Boat	79.5028	87.7162	35.5000	2.6226	2.8826
Goldhill	75.0487	119.8618	35.9000	2.6253	2.8137
Girl	99.5779	78.6989	35.7000	2.6302	3.3147

Respective simulations were also captured for CICC and M-MRSLS algorithms. A single run simulation was considered as an instance for all the test images for the proposed methods as shown in Figs. 4.11 and 4.12. These simulations present the converged PSNR value, number of learning attempts and elapsed value for each test image. Four different representations of candidates are considered in the graph of Fig. 4.11 for every test image. The representations were 'Asterisk', 'Circle', 'Square' and 'Diamond' for Candidate 1, Candidate 2, Candidate 3, and Candidate 4, respectively. As shown in Fig. 4.12, a mode value was considered for the graph of M-MRSLS. A random PSNR value was selected for each of the iterations. As there was a total of 40 iterations in a single run, there would be a total of 40 PSNR values, in which a mode value was considered.

Table 4.6 Comparison: computational time for the best, median and worst case

Images		Methods	
		CICC	M-MRSLS
Lena	Best	207.3164	41.2889
	Median	265.9226	259.7618
	Worst	520.5065	354.3885
Baboon	Best	312.6016	90.6541
	Median	633.4719	233.4110
	Worst	641.8447	467.0896
Woman	Best	185.9121	71.3670
	Median	238.2164	286.6590
	Worst	506.3748	485.2027
Boat	Best	248.9599	58.9555
	Median	453.5200	368.7303
	Worst	459.7481	368.7303
Goldhill	Best	234.4130	58.3548
	Median	379.5108	236.3064
	Worst	474.4410	500.6337
Girl	Best	188.2522	89.3751
	Median	323.2341	89.3751
	Worst	483.4530	379.3276

Table 4.7 Comparison: capacity (bits)

Capacity (bits)	Methods			
	CICC	M-MRSLS	JPEG_PSO	JQTM
Selected DCT coefficients for hiding	36	36	36	26
Number of bits to be hidden per DCT coefficient	2	2	2	2
Hiding capacity per block	72	72	72	52

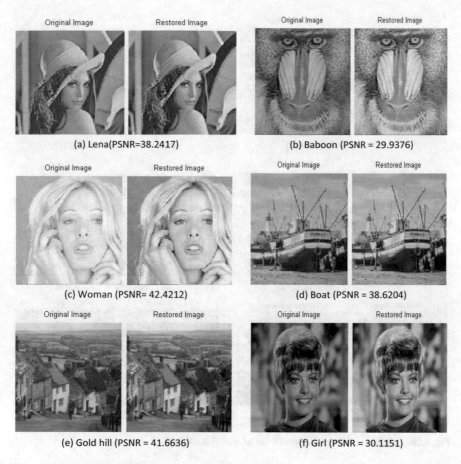

Fig. 4.9 Original and stego-images of the CICC method

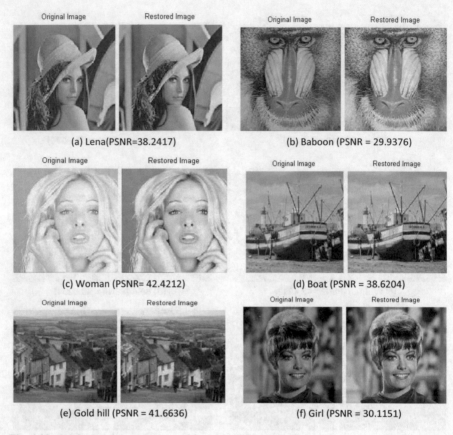

(a) Lena(PSNR=38.2417)   (b) Baboon (PSNR = 29.9376)

(c) Woman (PSNR= 42.4212)   (d) Boat (PSNR = 38.6204)

(e) Gold hill (PSNR = 41.6636) (f) Girl (PSNR = 30.1151)

Fig. 4.10   Original and stego-images of the M-MRSLS method

Simulations: CICC and M-MRSLS

1.Lena

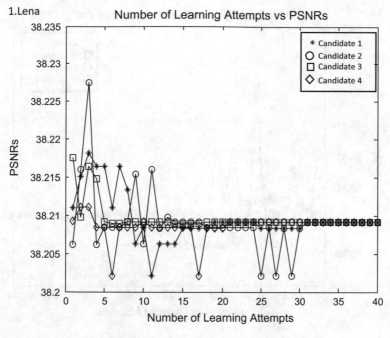

2.Baboon

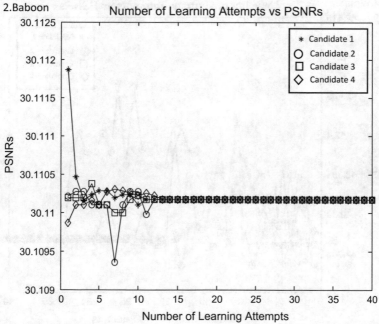

Fig. 4.11 CICC simulations (40 number of learning attempts)

3.Girl

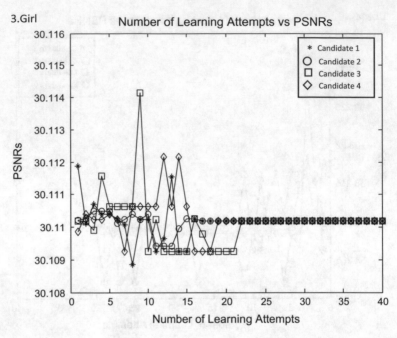

4.Woman

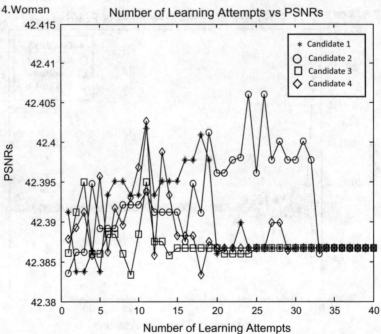

Fig. 4.11 (continued)

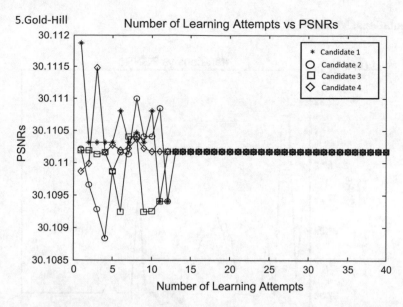

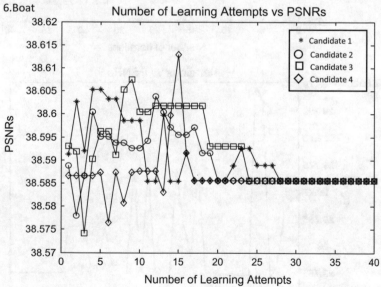

Fig. 4.11 (continued)

Simulations: M-MRSLS

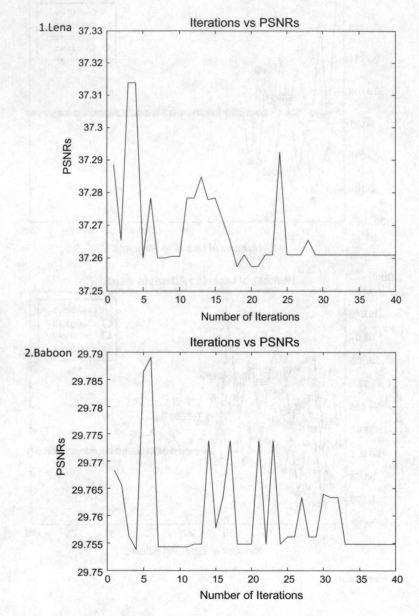

Fig. 4.12 M-MRSLS simulations (40 iterations)

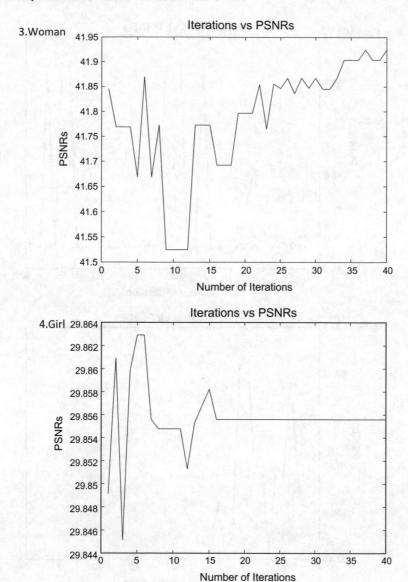

Fig. 4.12 (continued)

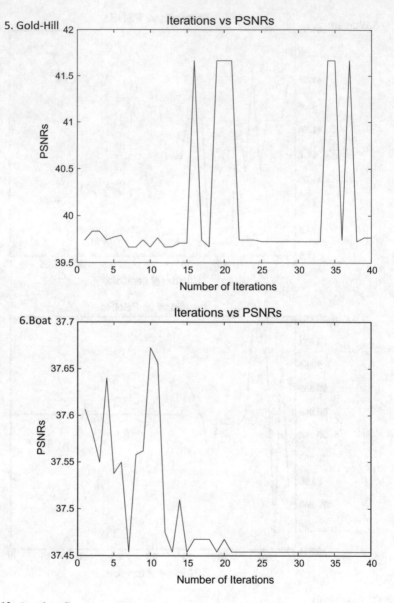

Fig. 4.12 (continued)

4.3 Summary

The methods CICC and M-MRSLS were applied in the Image processing domain and implemented by using six grayscale test images of size 256×256. The implementation of the proposed methods was done to resolve the two major concerns for image steganography i.e. capacity in bits and quality of the stego image. The proposed methods derive an optimal substitution matrix by CICC and M-MRSLS to transform the secret messages and then hide the secret message into the cover-image through a modified JPEG quantization table. The comparative results are discussed in Sect. 4.2 which shows that the solution of our methods with other cotemporary methods is comparable or even some times better than the other methods. The proposed methods achieved larger message capacity and better image quality than JQTM, Li et al. [6] and non-optimal substitution methods. Moreover, these methods provided high security to extract the secret message because, in order to recover the secret messages correctly, one should have the information about the optimal substitution matrix. Also, the PSNR values for the proposed methods were found better than the other mentioned methods for most of the test images. Therefore, the proposed methods provided better stego image quality, which made it difficult for the attacker to doubt on the system about the secret message and made the system secure. Though the proposed methods had many advantages, few limitations were also being observed. The computational time for the proposed methods was in the little higher side while comparing with other parallel methods which could be an area of improvement. The capacity of the embedded secret text could also be improved. Further, the robustness of the proposed methods could be observed and refined. These limitations are addressed in the next Chap. 5.

References

1. Chang, C.C., Chen, T.S., Chung, L.Z.: A steganographic method based upon JPEG and quantization table modification. Inf. Sci. **141**, 123–138 (2002)
2. Chen, Y., Argentinis, E., Weber, G.: IBM Watson: how cognitive computing can be applied to big data challenges in life sciences research. Clin. Ther. **38**(4), 688–701 (2016)
3. Kim, T.K., Paik, J.K.: Fast image restoration for reducing block artifacts based on adaptive constrained optimization. J. Vis. Commun. Image Represent. **9**(3), 234–242 (1998)
4. Kulkarni, O., Kulkarni, N., Kulkarni, A., Kakandikar, G.: Constrained cohort intelligence using static and dynamic penalty function approach for mechanical components design. Int. J. Parallel Emergent Distrib. Syst. **33**(6), 570–588 (2016)
5. Langelaar, G.C., Setyawan, I., Lagendijk, R.L.: Watermarking digital image and video data: a state-of-the-art overview. IEEE Signal Process. Mag. **17**(5), 20–46 (2000)
6. Li, X., Wang, J.: A steganographic method based upon JPEG and particle swarm optimization algorithm. Inf. Sci. **177**, 3099–3109 (2007)
7. Miano, J.: Compressed image file formats: JPEG, PNG, GIF, XBM, BMP, pp. 1–264. ACM Press Series, Addison-Wesley Professional, Boston (1999)
8. Sarmah, D., Kulkarni, A.J.: JPEG based steganography methods using cohort intelligence with cognitive computing and modified multi random start local search optimization algorithms. Inf. Sci. **430–431**, 378–396 (2018)
9. Wang, R.Z., Lin, C.F., Lin, J.C.: Image hiding by optimal LSB substitution and genetic algorithm. Pattern Recogn. **34**(3), 671–683 (2001)
10. Xu, G., Cui, Q., Shi, X., Ge, H., Zhan, Z., Lee, H.P., Liang, Y., Tai, R., Wu, C.: Particle swarm optimization based on dimensional learning strategy. Swarm Evolut. Comput. **45**, 33–51 (2019)
11. Zhu, F.: Blocking artifacts reduction in compressed data. In: Proceedings of the 2009 International Conference on Computer Engineering and Applications, IPCSIT, vol. 2. IACSIT Press, Singapore (2011)

Chapter 5
Secret Text Embedding Capacity Improvement (Approach 3)

In this chapter, the comprehensive procedure to improve the capacity of secret text is discussed. Two Steganography techniques were proposed which employed JPEG compression on grayscale image to hide secret text. In order to improve the capacity, a quantization table of size 16×16 [1] was preferred over a standard JPEG quantization table of size 8×8. In the proposed work, CICC (refer to Sect. 5.4) was applied to steganography to produce good results in terms of secret text embedding capacity and picture quality. Also, another motivating optimization algorithm referred to as MRSLS was modified to the M-MRSLS algorithm (refer to Sect. 5.4) and applied to steganography. Simulations were carried on six grayscale images for testing and validating with the comparable algorithms. The same test images were used as described in Sect. 4.2 to provide the same ground for comparison between Approach 3 and Approach 2. Experimental results revealed that this approach increases the embedding capacity of secret text without compromising the visual quality of stego images.

The different quantization tables of size 16×16 and its advantages and disadvantages are discussed in Sect. 5.1. The embedding and extraction procedure of steganography with CICC and M-MRSLS algorithms using a 16×16 quantization table are discussed in Sect. 5.2.

5.1 Quantization Table (Size: 16 × 16) for JPEG Image Compression

In this section, with the aim to improve the embedding capacity in the grayscale image, the cover image was divided into non-overlapping blocks of 16×16 pixels and used 16×16 optimal quantization table [1]. Researchers [3, 5] had already used this quantization table in a grayscale image; however, optimization techniques were not been applied yet in this combination. In order to explore further in this direction, an effort was made to implement two optimization techniques CICC and M-MRSLS

D. K. Sarmah et al., *Optimization Models in Steganography Using Metaheuristics*, Intelligent Systems Reference Library 187, https://doi.org/10.1007/978-3-030-42044-4_5

which were exploited to generate a ciphertext and an optimization matrix. The embedding of this optimized ciphertext was done in the DCT quantized values generated using 16×16 quantization tables. The following Section describes steganography with CICC and M-MRSLS algorithm using a 16×16 quantization table.

5.2 Steganography with CICC and M-MRSLS Algorithm Using 16 × 16 Quantization Table

In this approach, the proposed optimization algorithms referred to as CICC and M-MRSLS were used with a 16×16 quantization table to improve the embedding capacity of the secret text. The detail explanation of the embedding and extraction procedure with a block diagram and illustration is discussed in the following sections.

5.2.1 Embedding Procedure with Example

Four phases were considered in the embedding procedure as shown in Fig. 5.1: Grayscale image Pre-processing (Phase 1), Message Encryption and Optimal Matrix Identification (Phase 2), Secret Message insertion or Embedding (Phase 3), and Run Length Encoding (RLE), Digital Pulse Code Modification (DPCM), Entropy Coding (EC), Huffman Coding (HC), Frame Building (FB) and getting a Stego Image (Phase 4). CICC and M-MRSLS were employed to identify the optimal substitution matrix

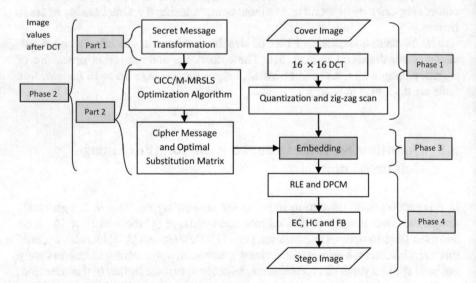

Fig. 5.1 Block diagram of the embedding procedure

and convert plain secret messages to cipher messages. An effort was put to solve the limitations of steganography by using a 16×16 quantization table. By implementing and applying two optimization algorithms in JPEG grayscale image, it enabled us to include the problem definition, objective function, and constraints.

Problem Definition: Increase the embedded capacity of secret messages and improve the security of the system by increasing the quality of the image.

Inputs: For the implementation of the two steganography methods three inputs were considered: grayscale cover image, secret message, and a number of substitution matrices.

Objectives: The following objectives were identified:

(i)   Encrypt the secret message and embed the encrypted secret message (cipher message) into the cover image to enhance the security level.
(ii) Improve PSNR value for the stego image.
(iii) Upgrade the cipher message capacity.

Figure 5.1 shows the flowchart for the whole procedure. Every step of the two steganography methods is discussed in its respective phases.

(i) *Phase 1 (Segmentation of Cover Image and Quantization)*: This was the pre-processing phase used to segment the grayscale cover image into 16×16 pixels of non-overlapping image blocks. Then DCT was applied to each block to transform its pixel value into DCT coefficients. DCT coefficients were further scaled using a 16×16 quantization table as shown in Table 5.1. In this quantization table, the positions where was used to embed the secret message. The quantized DCT coefficients were then rounded off to the nearest integers as similar to the JPEG image compression (refer Sect. 2.2). In order to select the coefficients for embedding, a traversal was done in the zigzag scan order. The table for zigzag scanning is shown in Table 5.2. There is a total of 256 values starting from 1 to 256 for a 16×16 image block matrix.

(ii) *Phase 2*: This phase describes the conversion of secret plain text to ciphertext. This was inspired by the OLSBS method [9]. This phase is divided into two parts. Part-1 explains the secret message transformation, Part-2 describes how the optimization algorithm was applied in the secret text and the optimal substitution matrix was selected.

(a) **Part-1 (Transformation of Secret Message)**

This part was considered the same as described in Sect. 4.1.1. The secret message was accepted and the optimization algorithm, either CICC or M-MRSLS was applied to achieve the optimized matrix and get the ciphertext. The accepted secret message was converted into a smaller number of bits. A decimal value d was selected to make individual groups having d number of bits. The decimal value d signifies the number of LSBs used from each DCT coefficient to hide d numbers of secret bits. The possible number of secret bit combinations is in the range from 0 to $2^d - 1$ signifies the range of decimal values. The value of d was assumed as 2 for easiness. Thus, the range of decimal values for secret bits was from 0 to 3. Substitution Matrix M (refer Eq. 4.1.1a) was considered to convert the plain secret bits to cipher secret

Table 5.1 16 × 16 quantization table [1]

7	7	7	7	7	1	1	1	1	1	1	1	1	1	1	1
7	7	7	7	1	1	1	1	1	1	1	1	1	1	1	17
7	7	7	1	1	1	1	1	1	1	1	1	1	1	17	18
7	7	1	1	1	1	1	1	1	1	1	1	1	17	18	20
7	1	1	1	1	1	1	1	1	1	1	1	17	18	20	22
1	1	1	1	1	1	1	1	1	1	1	17	18	20	22	24
1	1	1	1	1	1	1	1	1	1	17	18	20	22	24	26
1	1	1	1	1	1	1	1	1	17	18	20	22	24	26	28
1	1	1	1	1	1	1	1	17	18	20	22	24	26	28	30
1	1	1	1	1	1	1	17	18	20	22	24	26	28	30	33
1	1	1	1	1	1	17	18	20	22	24	26	28	30	33	36
1	1	1	1	1	17	18	20	22	24	26	28	30	33	36	39
1	1	1	1	17	18	20	22	24	26	28	30	33	36	39	42
1	1	1	17	18	20	22	24	26	28	30	33	36	39	42	45
1	1	17	18	20	22	24	26	28	30	33	36	39	42	45	49
1	17	18	20	22	24	26	28	30	33	36	39	42	45	49	52

Table 5.2 16 × 16 zigzag scanning

1	2	17	33	18	3	4	19	34	49	65	50	35	20	5	6
21	36	51	66	81	97	82	67	52	37	22	7	8	23	38	53
68	83	98	113	129	114	99	84	69	54	39	24	9	10	25	40
55	70	85	100	115	130	145	161	146	131	116	101	86	71	56	41
26	11	12	27	42	57	72	87	102	117	132	147	162	177	193	178
163	148	133	118	103	88	73	58	43	28	13	14	29	44	59	74
89	104	119	134	149	164	179	194	209	225	210	195	180	165	150	135
120	105	90	75	60	45	30	15	16	31	46	61	76	91	106	121
136	151	166	181	196	211	226	241	242	227	212	197	182	167	152	137
122	107	92	77	62	47	32	48	63	78	93	108	123	138	153	148
183	198	213	228	243	244	229	214	199	184	169	154	139	124	109	94
79	64	80	95	110	125	140	155	170	185	200	215	230	245	246	231
216	201	186	171	156	141	126	111	96	112	127	142	157	172	187	202
217	232	247	248	233	218	203	188	173	158	143	128	144	159	174	189
204	219	234	249	250	235	220	205	190	175	160	176	191	206	221	236
251	252	237	222	207	192	208	223	238	253	254	239	224	240	255	256

bits. Wang et al. [9] considered the substitution matrix in the form of an identity matrix and its variations. The total number of substitution matrices for the value d is $2^d!$ denoted as $M_1, \ldots, M_{2^{d!}}$. Thus, the total number of substitution matrices for the value $d = 2$ was 24 having a size of 4×4 of each M.

The important steps for this phase are as follows:

(i)   Accept the secret message.
(ii)   Convert the secret message into a number of bits.
(iii) Select d number of bits to make separate groups having d bits.
(iv)   Evaluate the decimal values of each individual group.
(v)   Generate the ciphertext by using substitution matrices.
(vi)   Apply the CICC algorithm or M-MRSLS algorithm to search for an optimal substitution matrix and its corresponding ciphertext.

A demonstration is shown below to transform the secret text into ciphertext with the help of the following steps:

Step-1: Let's consider a substitution matrix M.

$$M: \begin{array}{c} \\ 0 \\ 1 \\ 2 \\ 3 \end{array} \begin{array}{cccc} 0 & 1 & 2 & 3 \\ \begin{bmatrix} 0 & 0 & 1 & 0 \\ 1 & 0 & 0 & 0 \\ 0 & 1 & 0 & 0 \\ 0 & 0 & 0 & 1 \end{bmatrix} \end{array}$$

a.

Step-2: Assume the secret message bits are $\{10110001\}$. As the value of $d = 2$, divide the number of bits into a group of 2 bits. Now the secret message is

b.   Secret message: $\{10110001\}$

Step-3: Calculate the decimal value of each group of a secret message as present in Step 2 (b).

c.   Decimal values of secret message: $\{2\ 3\ 1\ 0\}$

Step-4: Replace the decimal values of the secret message which is considered as the row index of M by its respective column index where the value of M is found as 1.

d.   Substitution result of (c) using (a): $\{1\ 3\ 0\ 2\}$

Step-5: Convert the decimal values as shown in Step 4 (d) into a binary form, ensuing the ciphertext of a secret text using the substitution matrix M as mentioned in Step 2 and Step 1 respectively.

e.   The binary value of (d): $\{01\ 11\ 00\ 10\}$

The bits of the cipher message was hidden into the selected DCT coefficients as specified in Table 5.1. The decryption of a ciphertext was also be done by using transpose of M. As there was a total of 24 possible numbers of combinations of M for $d = 2$ and for each combination of M, the stego image quality might be different. Thus, there was a need to implement the optimization algorithm which could result in the optimal substitution matrix and its corresponding ciphertext.

(b) **Part 2 (Optimal Substitution Matrix Identification using CICC/M-MRSLS)**

This part is divided into two stages. Stage-1 discusses the identification of the optimal substitution matrix and its corresponding cipher message by using CICC whereas Stage-2 describes the M-MRSLS algorithm to identify the optimal substitution matrix and the corresponding cipher message. Illustrations are also shown for CICC and M-MRSLS algorithms in Stage-1 and Stage-2 respectively.

- *Stage-1: Optimal Substitution Matrix Identification using CICC*

As discussed before the need for the optimization algorithm to identify the optimal substitution matrix, a well-known optimization algorithm CI (for more details refer to Sects. 3.3 and 3.4), developed by Kulkarni et al. [6] was used along with the concept of cognitive computing (CC). The flowchart of CICC is shown in Fig. 4.2. Each substitution matrix M was considered as a candidate M_c in the cohort where $c = 1, 2, \ldots, C$. There was a total of 24 candidates for k = 2. In order to understand the behavior of each candidate and to get more clarity, a total of 4 candidates were selected randomly out of 24. Here behavior of any candidate referred to its quality which further helped to identify the behavior of a cohort (refer Sects. 3.3 and 3.4 for more detail). PSNR was considered as a fitness function to evaluate the quality of the stego image. The position of a row of substitution matrix M having value 1 decided the quality of a candidate. Thus, each candidate possessed 4 qualities due to the size of M i.e. 4×4.

The considered runs for implementation were 20 to analyze the results. Since the overall behavior of the cohort may improve after each and every iteration thus the maximum numbers of iterations under each run were selected 40 and the threshold limit was set as 0.0001. The steps, flowchart and a sample illustration are shown below to describe the overall procedure of each stage.

Step 1: Initialize the cohort with a random number of candidates where any candidate is represented as M_c, $c = 1, 2, \ldots, C$.

Step 2: Evaluate the *PSNR* value for each candidate M_c with respect to the secret message.

Step 3: Evaluate the selection probability P^{M_c} of any candidate M_c as follows:

$$P^{M_c} = \frac{PSNR_{M_c}}{\sum_{c=1}^{C} PSNR_{M_c}} \tag{5.2.1a}$$

Step 4: Apply a roulette wheel algorithm by any candidate M_c, $c = 1, 2, \ldots, C$ to select and follow a candidate as per the results generated by the algorithm. Here, the term 'follows' refers to the quality adapted by the follower candidate to the followed candidate. Each candidate generated a random integer from within 1 to 2^d The random integer decided the row value of the follower candidate to be replaced with the corresponding row value of the followed candidate.

Step 5:   Apply cognitive computing (CC) approach with CI to obtain a better
          solution. Experimentation showed since CI was based on a probabilistic
          approach that did not guarantee to have convergence for a candidate's PSNR
          value. Thus, the concept of CC was used along with CI which enabled each
          candidate to accept a better quality and improved the overall behavior of
          the cohort. According to the CICC approach, if the candidate's PSNR in
          the current iteration is better than the PSNR of the previous iteration then
          accepting the change else retaining earlier.
Step 6:   Execute the conditions concurrently as described below to achieve the cohort
          saturation:

          (a)   If the maximum number of learning efforts are reached.
          (b)   If the cohort does not improve its behavior after a certain number of runs
                i.e. there is no considerable change or difference identified between the
                PSNR values of all the candidates in the continuous learning efforts.

The flowchart of CICC is shown in Fig. 5.2 and its whole process is explained
through a sample illustration presented in Fig. 5.3.

- *Stage-I: A sample Illustration of a grayscale image using CICC*

Total 4 candidates were selected from the given number of candidates M_c, $c =$
$1, 2, \ldots, 2^d$ as the value of d was considered 2. A grayscale image of Woman having
size 256×256 is considered for this illustration. A single iteration is presented in
Fig. 5.3.

- *Stage-2: Optimal Substitution Matrix Identification using M-MRSLS*

The original algorithm MRSLS was developed by Kulkarni et al. [4] and is explained
in detail in Sect. 3.2.3. MRSLS algorithm was modified, implemented and applied
on a JPEG grayscale image having 16×16 quantized coefficients for steganography.
This algorithm is referred to as M-MRSLS. The flowchart and its illustration are
shown in Figs. 5.4 and 5.5 respectively. In the original algorithm of MRSLS (refer to
Sect. 3.2.3), a duo swapping approach is used which enable every solution in the set
M_p to generate a neighboring solution, whereas the M-MRSLS algorithm generated
a random solution that was dependent upon the associated solution's behavior. The
number of substitution matrices as described in Part 1, were considered as the number
of solutions for a set $\{M_p, \ p = \{1, 2, 3 \ldots, P)$ where $P = 2^d!\}$. Since d value was
selected as 2, hence there was a total of 24 solutions. The total number of selected
solutions was 4 in order to get more clarity and easiness. These solutions were
picked up randomly. A fitness function i.e. PSNR, as described in Eq. 4.1.1b, was
considered to determine the quality of a stego image which is dependent upon the
behavior of a set M_p. The complete performance of the set could be determined
by the overall performance/behavior of every solution. The quality of each solution
makes its behavior. In M_p, the row position having value 1 was considered the quality
of a solution. The size of each solution which is in the form of substitution matrix
was taken 4×4, which implies 4 qualities for every solution in the set. The same
number of runs and same iterations under each run were considered for M-MRSLS

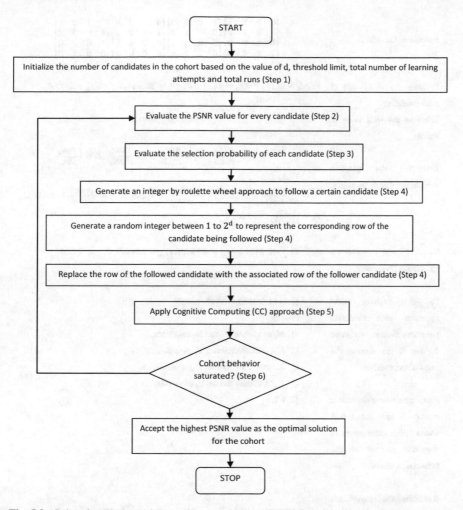

Fig. 5.2 Cohort intelligence with cognitive computing (CICC) flowchart (approach 3)

as discussed in Stage-1 for CICC i.e. 20 runs and a maximum of 40 iterations under each run. The steps in detail are discussed below.

Step 1:   Produce the P random solutions for a set wherein the representation of any solution is $\{M_p, \; p = \{1, 2, 3 \ldots, P) \text{ where } P = 2^d!\}$

Step 2:   Evaluate the PSNR value for each solution M_p. Equation-1 is used to determining the PSNR value.

Number of Candidates

$$
\begin{bmatrix} 0 & 0 & 0 & 1 \\ 0 & 0 & 1 & 0 \\ 0 & 1 & 0 & 0 \\ 1 & 0 & 0 & 0 \end{bmatrix}
\begin{bmatrix} 1 & 0 & 0 & 0 \\ 0 & 0 & 1 & 0 \\ 0 & 0 & 0 & 1 \\ 0 & 1 & 0 & 0 \end{bmatrix}
\begin{bmatrix} 0 & 1 & 0 & 0 \\ 1 & 0 & 0 & 0 \\ 0 & 0 & 1 & 0 \\ 0 & 0 & 0 & 1 \end{bmatrix}
\begin{bmatrix} 0 & 1 & 0 & 0 \\ 0 & 0 & 0 & 1 \\ 1 & 0 & 0 & 0 \\ 0 & 0 & 1 & 0 \end{bmatrix}
$$

a) M_1 b)M_2 c) M_3 d) M_4

Evaluate the PSNR value of each candidate

$PSNR_1 = 38.353, PSNR_2 = 38.3113, PSNR_3 = 38.3719, and PSNR_4 = 38.3797$

Calculate the total value of PSNR

$$PSNR_T = 153.4159$$

Determine the selection probability of each candidate.

$$P_1 = \frac{PSNR_1}{PSNR_T}, P_2 = \frac{PSNR_2}{PSNR_T}, P_3 = \frac{PSNR_3}{PSNR_T}, P_4 = \frac{PSNR_4}{PSNR_T}$$

Thus,

$P_1 = \frac{38.353}{153.4159} = 0.24999, \ P_2 = \frac{38.3113}{153.4159} = 0.24972, P_3 = \frac{38.3719}{153.4159} = 0.25012,$

$P_4 = \frac{38.3797}{153.4159} = 0.25017$

Calculate the candidate's cumulative probability.

$P_{1_{cu}} = 0.24999, P_{2_{cu}} = 0.49971, P_{3_{cu}} = 0.74983, P_{4_{cu}} = 1.000$

By using roulette wheel approach, each candidate generates 4values between 0 and 1 to follow the convincing candidate.

$\{0.4444, 0.8923, 0.5649, 0.1445\}$

Implies

a) M_1 candidate follows M_2 candidate,

b) M_2candidate follows M_4 candidate,

c) M_3 candidate follows M_3candidate, and

d) M_4Candidate followsM_1 candidate.

Every candidate generates a random integer between 1 and 4 which corresponds to the row number of the followed candidate.

$\{3, 4, 1, 2\}$

Substitute the row$_3$of the candidate M_1 with row$_3$of the candidate M_2 , row$_4$ of the candidate M_2 with row$_4$ of the candidate M_4, row$_1$ of

Candidate M_1: $\begin{bmatrix} 0 & 0 & 0 & 1 \\ 0 & 0 & 1 & 0 \\ 0 & 1 & 0 & 0 \\ 1 & 0 & 0 & 0 \end{bmatrix}$ $row_3$of the candidate M_1: $\{0\ 1\ 0\ 0\}$

Candidate M_2: $\begin{bmatrix} 1 & 0 & 0 & 0 \\ 0 & 0 & 1 & 0 \\ 0 & 0 & 0 & 1 \\ 0 & 1 & 0 & 0 \end{bmatrix}$ $row_3$of the candidate M_2: $\{0\ 0\ 0\ 1\}$

Fig. 5.3 Illustration of a grayscale image using CICC

the candidate M_3 with row_1 of the candidate M_3 and row_2 of the candidate M_4 with row_2 of the candidate M_1 respectively to adapt the quality of the followed candidate by the follower candidate	After replacement, the new candidate M_1' : $\begin{bmatrix} 0 & 0 & 0 & 1 \\ 0 & 0 & 1 & 0 \\ 0 & 0 & 0 & 1 \\ 1 & 0 & 0 & 0 \end{bmatrix}$
The new 4 candidates in the cohort	$\begin{bmatrix} 0 & 0 & 0 & 1 \\ 0 & 0 & 1 & 0 \\ 0 & 0 & 0 & 1 \\ 1 & 0 & 0 & 0 \end{bmatrix}$ $\begin{bmatrix} 1 & 0 & 0 & 0 \\ 0 & 0 & 1 & 0 \\ 0 & 0 & 0 & 1 \\ 0 & 0 & 1 & 0 \end{bmatrix}$ $\begin{bmatrix} 0 & 1 & 0 & 0 \\ 1 & 0 & 0 & 0 \\ 0 & 0 & 1 & 0 \\ 0 & 0 & 0 & 1 \end{bmatrix}$ $\begin{bmatrix} 0 & 1 & 0 & 0 \\ 0 & 0 & 1 & 0 \\ 1 & 0 & 0 & 0 \\ 0 & 0 & 1 & 0 \end{bmatrix}$ (a) M_1' (b) M_2' (c) M_3' (d) M_4'
Calculate the PSNR values of the 4 new candidates	$PSNR'_1 = 38.1833, PSNR'_2 = 38.1091, PSNR'_3 = 38.2017, and PSNR'_4 = 38.3937$
Apply CC concept in every PSNR values of all the candidates and compare these values from the current to the previous iteration.	$M_1' < M_1$ $M_2' < M_2$ $M_3' < M_3$ $M_4' > M_4$
Accept the high PSNR values for every candidate and consider it as the current iteration value which enables the cohort to reach its optimal value.	$39.1833, 38.2091, 38.2417, 38.2037$

Continue the process until the saturation condition is reached as mentioned in **Step 6** of Part 2 under Stage 1.

Fig. 5.3 (continued)

Step 3: Every solution M_p, $p = 1, 2, \ldots, P$ generates a random integer between 1 and p and follows a solution associated with the generated integer. For example, assume that each solution M_p, $p = 1, 2, \ldots, P$ generates a random integer 4 implies that each solution of the set follows the solution M_4.

Step 4: In order to follow a certain solution as described in Step 3, each solution generates further a random integer from within 1 to 2^d. This integer represents the corresponding row of the solution being followed. To adapt the qualities of the followed solution, the values of this row number replaces the values of the associated row of the follower solution.

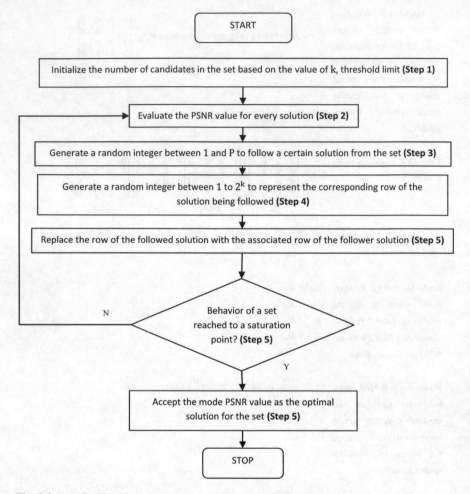

Fig. 5.4   Modified Multi random start local search (M-MRSLS) flow chart (approach 3)

Step 5:   The goal of this algorithm is to determine an optimized fitness solution amongst the generated feasible solutions. Since there is great randomization involved in the M-MRSLS algorithm, an idea of statistical Mode is incorporated in the algorithm. The Mode is the number that occurs most frequently within a set of numbers. Mode helps identify the most common or frequent occurrence of a characteristic. The saturation point is considered as the maximum number of attempts i.e. 40.

Step 6:   All the steps are repeated until the saturation point is reached.

The Flowchart and mathematical illustration for M-MRSLS are shown in Figs. 5.4 and 5.5. An optimization matrix of size 4×4, from a set of an identity matrix and its variations, was obtained after implementation of either Stage-1 or Stage-2. The

Solutions

$$
a) \ M_1 \quad
\begin{bmatrix} 0 & 1 & 0 & 0 \\ 1 & 0 & 0 & 0 \\ 0 & 0 & 1 & 0 \\ 0 & 0 & 0 & 1 \end{bmatrix} \quad
b) M_2 \quad
\begin{bmatrix} 1 & 0 & 0 & 0 \\ 0 & 0 & 1 & 0 \\ 0 & 1 & 0 & 0 \\ 0 & 0 & 0 & 1 \end{bmatrix} \quad
c) M_3 \quad
\begin{bmatrix} 0 & 0 & 0 & 1 \\ 0 & 1 & 0 & 0 \\ 1 & 0 & 0 & 0 \\ 0 & 0 & 1 & 0 \end{bmatrix} \quad
d) M_4 \quad
\begin{bmatrix} 0 & 0 & 1 & 0 \\ 0 & 0 & 0 & 1 \\ 1 & 0 & 0 & 0 \\ 0 & 1 & 0 & 0 \end{bmatrix}
$$

Evaluate the $PSNR$ value of each solution	$PSNR_1 = 38.3476, PSNR_2 = 38.3710, PSNR_3 = 38.3638 \ and \ PSNR_4 = 38.3651$
Select a random value of PSNR	38.3710
Set generates a single random value between 1 and 4 to follow a certain solution.	2 Implies $\{M_1, M_2, M_3, M_4\}$ follows a solution M_2
Every solution generates a random integer between 1 and 4 which corresponds to the row number of the followed solution.	$\{3, 4, 1, 2\}$

Replace the row_3 of the solution M_1 with row_3 of the solution M_2, row_4 of the solution M_2 with row_4 of the solution M_2, row_1 of the solution M_3 with row_1 of the solution M_2 and row_2 of the solution M_4 with row_2 of the solution M_2 in order to adapt the quality of the followed solution by the follower solution

Solution M_1: $\begin{bmatrix} 0 & 1 & 0 & 0 \\ 1 & 0 & 0 & 0 \\ 0 & 0 & 1 & 0 \\ 0 & 0 & 0 & 1 \end{bmatrix}$ row_3 of the solution M_1: $\{0 \ 0 \ 1 \ 0\}$

Solution M_3: $\begin{bmatrix} 0 & 0 & 0 & 1 \\ 0 & 1 & 0 & 0 \\ 1 & 0 & 0 & 0 \\ 0 & 0 & 1 & 0 \end{bmatrix}$ row_3 of the solution M_3: $\{1 \ 0 \ 0 \ 0\}$

After replacement, the new solution $M_1{}'$: $\begin{bmatrix} 0 & 1 & 0 & 0 \\ 1 & 0 & 0 & 0 \\ 1 & 0 & 0 & 0 \\ 0 & 0 & 0 & 1 \end{bmatrix}$

The new 4 solutions in the set

$$
(a) \ M_1{}' \quad
\begin{bmatrix} 0 & 1 & 0 & 0 \\ 1 & 0 & 0 & 0 \\ 1 & 0 & 0 & 0 \\ 0 & 0 & 0 & 1 \end{bmatrix} \quad
(b) M_2{}' \quad
\begin{bmatrix} 0 & 0 & 0 & 1 \\ 0 & 1 & 0 & 0 \\ 1 & 0 & 0 & 0 \\ 0 & 0 & 1 & 0 \end{bmatrix} \quad
(c) M_3{}' \quad
\begin{bmatrix} 0 & 0 & 0 & 1 \\ 0 & 1 & 0 & 0 \\ 1 & 0 & 0 & 0 \\ 0 & 0 & 1 & 0 \end{bmatrix} \quad
(d) M_4{}' \quad
\begin{bmatrix} 0 & 0 & 1 & 0 \\ 0 & 1 & 0 & 0 \\ 1 & 0 & 0 & 0 \\ 0 & 1 & 0 & 0 \end{bmatrix}
$$

Calculate the PSNR values of the 4 new solutions	$PSNR'_1 = 38.3650, PSNR'_2 = 38.1091, PSNR'_3 = 38.3697, and PSNR'_4 = 38.3669$
Select a random value of PSNR	38.3697

Continue the process until the saturation condition is reached as mentioned in Step6 of <u>Stage 2</u>.

| Selected random PSNR values from each iteration | $38.3476, 38.3710, 38.3710, 38.3710, 38.3638, 38.3345, 38.3638, 38.3651,$
 $38.3690, 38.3690, 38.3690, \ 38.3707, 38.3707, 38.3621, 38.3621, 38.3650,$
 $38.3650, 38.3650, 38.3697, 38.3697, 38.3697, 38.3669, 38.3705, 38.3675$
 $38.3609, 38.3536, 38.3608, 38.3671, 38.3621, \ 38.3671, 38.3530, 38.3530$
 $38.3536, 38.3593, 38.3593, 38.3593, 38.3719, 38.3593, 38.3593, 38.3719$ |
| Accept the mode PSNR value | 38.3593 |

Fig. 5.5 Illustration a grayscale image using M-MRSLS

embedding procedure of the transformed secret message into the image blocks is described in Phase 3 following the illustration of the M-MRSLS algorithm.

- *Stage-2: A sample Illustration of a grayscale image using M-MRSLS*

The total number of solutions in a set M_p, $p = 1, 2, \ldots\ldots, 2^d$ for d = 2 is 4. A single iteration is shown considering a grayscale image Woman having size 256×256.

(iii) *Phase 3 (Embedding procedure of transformed secret text using 16×16 Quantization Table)*

As described earlier the generation of an optimal substitution matrix by applying either CICC algorithm or M-MRSLS algorithm, which in turn generated the transformed secret text, was used to hide into the low-mid frequency components of the quantized DCT coefficients. Any image was firstly divided into blocks and once the DCT was applied to each block, DCT coefficients were generated (refer Sect. 2.2). Each block of an image was fragmented into three frequency bands i.e. low, mid and high. The topmost left value of any image block was considered as the DC coefficient and the remaining coefficients were called AC coefficients. AC coefficients were used to hide the secret text as there could be a visual distortion of the image quality if the DC coefficients were altered [10]. Low to the middle-frequency zone of the coefficients were considered to be the safer one since the higher frequency zone was easily targeted by an attacker [7] and there was a probable chance of revealing the secret information by applying compression techniques and noise attacks. As described in Sect. 2.2.1 of Chap. 2 (the overall procedure of JPEG), hiding of transformed secret text was done after quantization. As described in the previous section, the quantization table plays a very major role for quantization as a number of quantized coefficients would be selected as per the size and the values of the quantization table. The embedding procedure of transformed secret text is shown in Fig. 5.6.

(iv) *Phase 4 (Generation of Stego Image)*

This phase includes five sub-phases RLE, DPCM, EC, HC and FB which were applied to the embedded output (refer Fig. 5.7) to generate the stego image. These sub-phases are explained in detail in Sect. 2.2. RLE and DPCM were applied only on AC components and DC components of the embedded output respectively. The produced output from the previous phase was used for EC in which DC, as well as AC components, were encoded. Then, HC and FB were applied to these generated encoded values to produce a stego image. All the applied processes on embedded output under this phase were used for the encoding purpose. The flowchart for this phase is shown in Fig. 5.7.

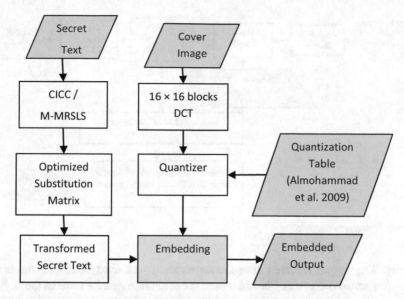

Fig. 5.6 Embedding procedure of CICC/M-MRSLS (approach-3)

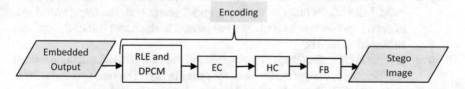

Fig. 5.7 Generation of stego image

5.2.2 Extraction Procedure

This algorithm describes the extraction procedure of secret text and cover image at the receiver end. The reverse process of the embedding algorithm is considered in this section which is represented in Fig. 5.8.

The whole procedure is divided into five steps as follows.

Step 1: The overall encoding procedure by different processes is shown in Fig. 5.7. The reverse procedure of encoding called decoding was applied to stego image which helped to retrieve the decoded blocks of stego image of size 16 × 16.

Step 2: Dequantization was applied to the produced output from **Step1**. The same quantization table as proposed by Almohammad et al. [1] was used for dequantization.

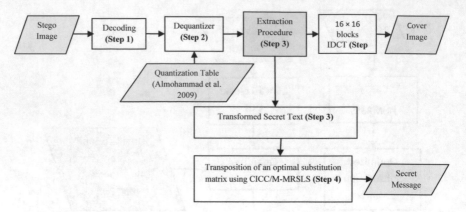

Fig. 5.8 Extraction procedure

Step 3:   The output received after Dequantization was used to extract the trans-
          formed secret bits. The order used for extraction was the same as the embed-
          ding order of the transformed secret text. The embedding order table is
          discussed and presented in Table 5.2.
Step 4:   Transposition of an optimal substitution matrix was done either using CICC
          or M-MRSLS. In order to get the original secret text, the transformed bits
          of secret text (output of Step 3) were passed to the transpose of the optimal
          substitution matrix.
Step 5:   Inverse DCT (refer Chap. 2) was applied to the 16×16 blocks of dequantized
          coefficients to extract the block of the cover image. The same procedure was
          repeated until all the blocks of the cover image were retrieved.

5.3   Comparison and Discussion on Results

This section shows the results of the proposed CICC and MRSLS. In order to implement the proposed methods, a total of six grayscale images were considered for testing
purposes. These six images were all 8-bit images having 256 gray levels and the size
of these images was taken as 256×256. The images considered for experimentation were Lena, Baboon, Boat, Gold Hill, Girl, and Women. The analysis of these
results was done in terms of evaluation parameters such as PSNR, computational
time, secret text capacity and a number of function evaluations. Figure 5.9 presents
the computed PSNR with respect to each image for both the proposed methods i.e.
CICC and M-MRSLS for the cover images, as well as stego images.

Steganography implemented on various Images

- Baboon Image 256 × 256 *Pixels*

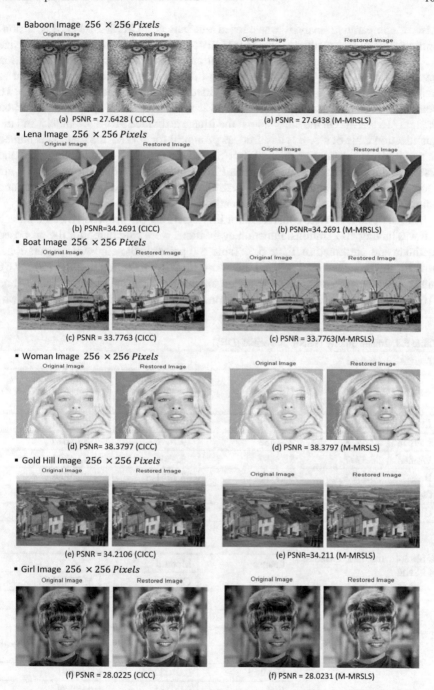

(a) PSNR = 27.6428 (CICC)   (a) PSNR = 27.6438 (M-MRSLS)

- Lena Image 256 × 256 *Pixels*

(b) PSNR=34.2691 (CICC)   (b) PSNR=34.2691 (M-MRSLS)

- Boat Image 256 × 256 *Pixels*

(c) PSNR = 33.7763 (CICC)   (c) PSNR = 33.7763(M-MRSLS)

- Woman Image 256 × 256 *Pixels*

(d) PSNR= 38.3797 (CICC)   (d) PSNR = 38.3797 (M-MRSLS)

- Gold Hill Image 256 × 256 *Pixels*

(e) PSNR = 34.2106 (CICC)   (e) PSNR=34.211 (M-MRSLS)

- Girl Image 256 × 256 *Pixels*

(f) PSNR = 28.0225 (CICC)   (f) PSNR = 28.0231 (M-MRSLS)

Fig. 5.9 Original and Stego-images of the CICC and M-MRSLS method

Also, a comparative analysis was carried out between CICC, M-MRSLS, a non-optimal substitution method of size 16×16, a non-optimal substitution method of size 8×8 and JQTM. The non-optimal substitution method was the naming convention given by Li and Wang [8] as there was no optimization methodology involved to transform the secret bits. If the image is divided into 16×16 blocks and the 16×16 quantization table is used then this method is named as a non-optimal substitution method (16×16). On the other side, if the image is divided into 8×8 blocks and the quantization table of size 8×8 is used, as proposed for JQTM, the method is named as non-optimal substitution method (8×8). All the methods were implemented and tested on the same images. Its results as PSNR and standard deviation of PSNR are shown in Tables 5.3 and 5.4. Image qualities of the proposed methods were found comparable with the other mentioned methods.

A total of 20 runs were considered for implementing the proposed methods. Each run was having 40 iterations. Function evaluations were calculated for the proposed methods and presented in Table 5.5. Table 5.6 presents the elapsed/computational time for the proposed methods and the non-optimal substitution method (16×16). Since there was no optimality included in the non-optimal substitution method (16×16) and was executed only one time, the evaluated computational time was

Table 5.3 Image quality [PSNR in Decibel (DB)]

Images	Method			
	CICC (16×16)	M-MRSLS (16×16)	Non optimal substitution (16×16)	Non optimal substitution (8×8)
Lena	34.2691	34.2691	37.7849	37.7840
Baboon	27.6428	27.6438	29.1646	29.1646
Woman	38.3797	38.3797	38.3546	42.1203
Boat	33.7763	33.7763	33.7541	38.5884
Goldhill	34.2106	34.2110	41.0410	41.0400
Girl	28.0225	28.0231	28.0174	29.5103

Table 5.4 Standard deviation of PSNR

Images	Method	
	CICC	M-MRSLS
Lena	0.01599	0.02550
Baboon	0.00219	0.00171
Woman	0.00344	0.02533
Boat	0.00509	0.01435
Goldhill	0.01082	0.01091
Girl	0.00181	0.00554

Table 5.5 Comparison of function evaluations (FE)

Images	Method	
	CICC	M-MRSLS
Lena	27	25
Baboon	32	36
Woman	31	22
Boat	32	39
Goldhill	24	35
Girl	24	18

Table 5.6 Comparison of computational time

Images	Method		
	CICC	M-MRSLS	Non-optimal substitution (16×16)
Lena	71.7879	81.6882	7.0765
Baboon	144.6961	121.876	8.8272
Woman	63.2178	92.5372	3.9194
Boat	115.796	102.5778	4.3911
Goldhill	108.9902	77.5181	7.7045
Girl	59.8623	84.2666	3.442

less than the proposed methods. Best, Median and worst-case PSNR values were evaluated for all the images of the proposed methods and shown in Tables 5.7 and 5.8 respectively. Tables 5.9 and 5.10 present the computational time for the proposed methods for best, median and the worst case. This computational time was evaluated for all the test images. The capacity of embedded secret text in terms of the number of bits was also calculated for the proposed methods and was compared with the non-optimal substitution method (8×8). The comparative analysis for the capacity is shown in Table 5.11. Since the proposed methods used the (16×16) quantization table, there was a total of 121 DCT coefficients selected in each block and 2bits were used per coefficient to hide the secret text. Thus, the total number of secret bits that could be embedded in each block was $121 \times 2 = 242$. The size of the cover image was considered 256×256. Therefore, the total numbers of image blocks were (256×256) $\div$ (16×16) $= 256$ which enables us to calculate the total embedding capacity of secret text in the image i.e. $256 \times 242 = 61952$ bits. However, this analysis was compared with non-optimal substitution method (8×8), which was implemented on the same images and used (8×8) quantization table as proposed in JQTM (total 26 DCT coefficients were selected and 2bits were used per coefficient). A total of $26 \times 2 = 52$ coefficients

Table 5.7 Comparison: best, median and worst Image quality (PSNR in DB) for Lena, Baboon and Woman

Method	Images								
	Lena			Baboon			Woman		
	Best	Median	Worst	Best	Median	Worst	Best	Median	Worst
CICC	34.2691	34.2585	34.2083	27.6438	27.6398	27.6371	38.3797	38.3719	38.3708
M-MRSLS	34.2691	34.2543	34.2083	27.6428	27.6398	27.6371	38.3797	38.35335	38.3113

Table 5.8 Comparison: best, median and worst Image quality (PSNR in DB) for Boat, Gold hill and Girl

Method	Images								
	Boat			Gold hill			Girl		
	Best	Median	Worst	Best	Median	Worst	Best	Median	Worst
CICC	33.7763	33.7652	33.7652	34.2106	34.1877	34.1754	28.0231	28.0186	28.018
M-MF-SLS	33.7763	33.75595	33.7361	34.2106	34.1877	34.1754	28.0225	28.0182	28.0087

Table 5.9 Comparison: computational time for the best, median and worst case for Lena, Baboon and Woman

Method	Images											
	Lena			Baboon			Woman					
	Best	Median	Worst	Best	Median	Worst	Best	Median	Worst			
CICC	207.4928	369.5112	484.7685	193.0027	454.4547	716.9827	252.8656	479.9283	533.0646			
M-MRSLS	78.9396	154.6788	316.5477	136.4402	289.1447	521.5855	46.109	237.9319	389.6469			

Table 5.10 Comparison: computational time for the best, median and worst case for Boat, Gold hill and Girl

Method	Images								
	Boat			Gold hill			Girl		
	Best	Median	Worst	Best	Median	Worst	Best	Median	Worst
CICC	222.3972	373.1209	527.4827	273.6345	479.6342	525.1040	272.3617	489.6918	564.7522
M-MRSLS	50.5418	204.1573	439.7497	49.8291	170.9784	309.8737	221.9242	298.2659	512.2002

Table 5.11 Comparison of capacity (bits)

Capacity (bits)	Methods		
	CICC	M-MRSLS	Non-optimal substitution method (8 × 8)
Selected DCT coefficients for hiding	121	121	26
Number of bits to be hidden per DCT coefficient	2	2	2
Hiding capacity per block	242	242	52
Total blocks for 256 × 256 image	256	256	1024
Total hiding capacity	61,952	61,952	53,248

were selected in each block which further calculated the total number of bits/capacity for all blocks/entire image i.e. Total number of blocks × 52. In this case, the total number of blocks was $(56 \times 256) \div (8 \times 8) = 1024$, hence the calculated capacity was $1024 \times 52 = 53248$. Hence, the capacity of the proposed methods with the (8×8) quantization table was less than the (16×16) quantization table. The increased percentage of capacity improvement was $((61952 - 53248) \div 53248) \times 100 = 16.35\%$. Thus it can be concluded that the secret text embedding capacity of the proposed methods is improved than the other methods selected for comparison.

Snapshots of the proposed methods are also presented to show the converged PSNR value, the number of learning attempts and elapsed value for each sample image. Figures 5.10 and 5.11 show the graph and the converged PSNR value tested for all the images for CICC and M-MRSLS respectively. A single run is showcased in these tables. Four different representations of candidates are shown in the graph of Fig. 5.10 for Lena's image. These representations are 'Asterisk', 'Circle', 'Square' and 'Diamond' for the Canidate1, Candidate2, Candidate3, and candidate4 respectively. The dimensions of the graph in x and y directions are represented as the number of learning attempts and PSNR respectively. Also, the other method i.e. M-MRSLS was analyzed as shown in Fig. 5.11, wherein a mode value was calculated based on a selected random PSNR for each iteration. There were 40 PSNRs for all the iterations under each run on which the mode value was considered. The Lena image was referred from Fig. 5.11. A single random value of PSNR was selected for each of the iterations; the representation of the candidate is shown in one shape i.e. 'Circle' in the graph.

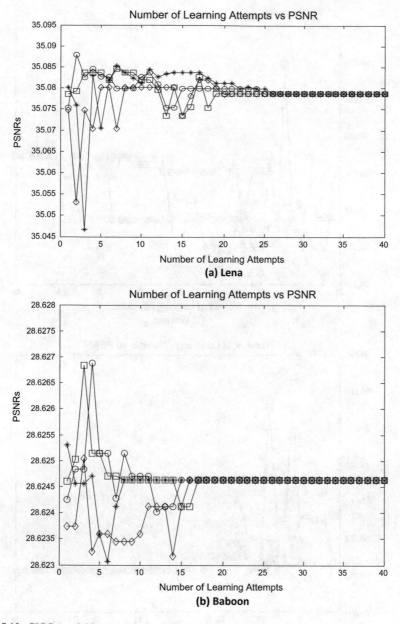

Fig. 5.10 CICC (total-20 runs) (approach 3)

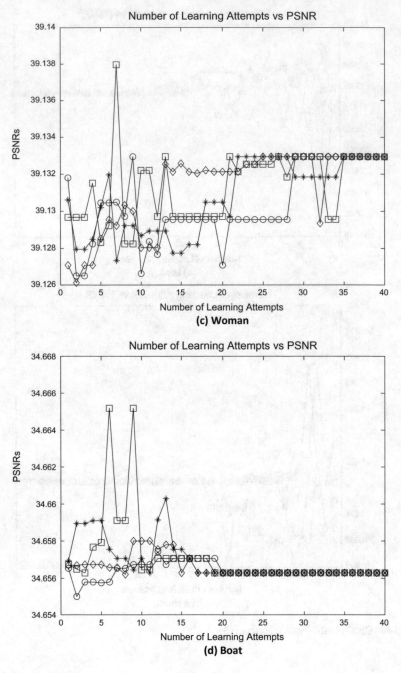

Fig. 5.10 (continued)

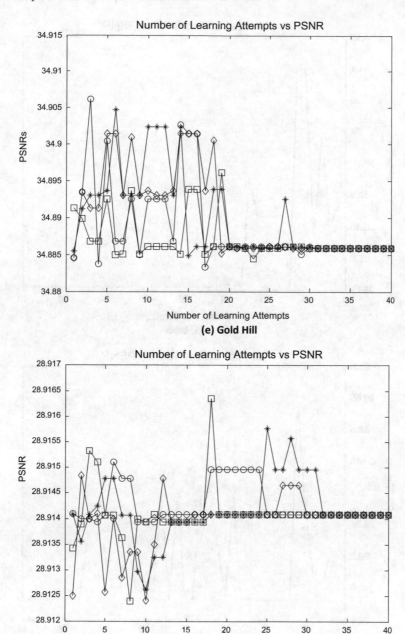

Fig. 5.10 (continued)

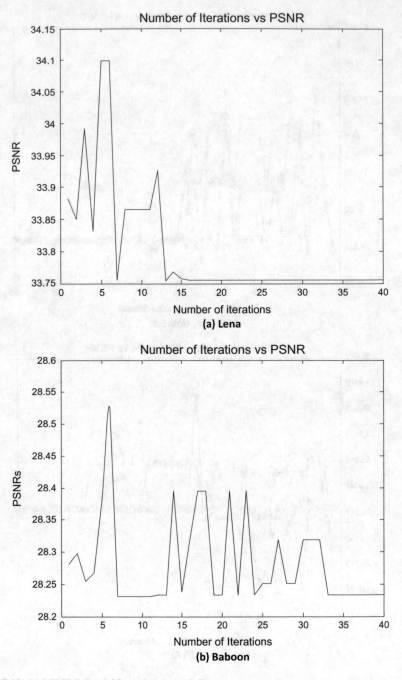

Fig. 5.11 M-MRSLS (total-20 runs) (approach 3)

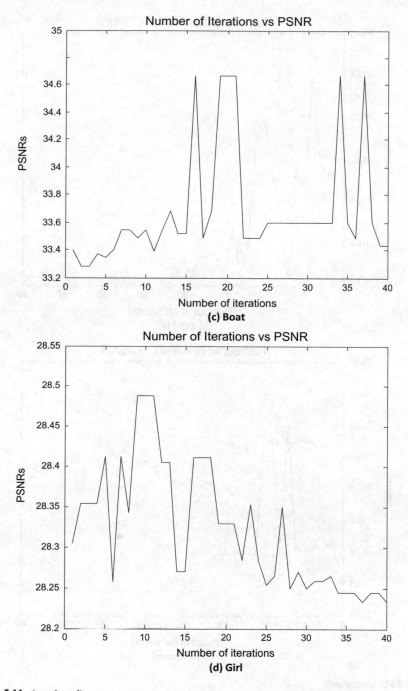

(c) Boat

(d) Girl

Fig. 5.11 (continued)

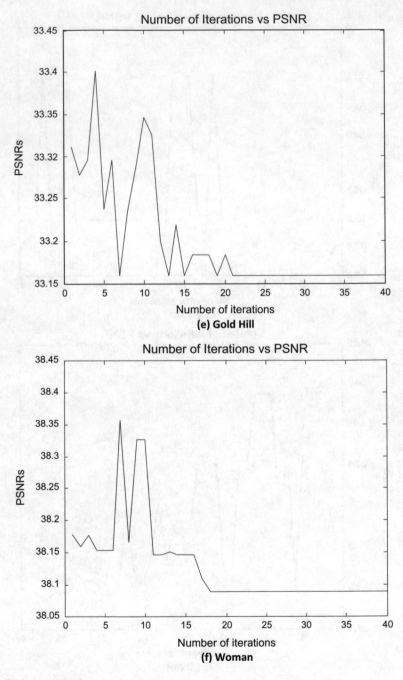

(e) Gold Hill

(f) Woman

Fig. 5.11 (continued)

5.4   Summary

Two secure JPEG steganography algorithms based on CICC and M-MRSLS optimization techniques were proposed and applied to six grayscale test images. Also, an effort was made to explore the said methods using a 16×16 quantization table. Since the JPEG steganography was used where the DCT calculation required for 8×8 pixels block. If the number of blocks gets increased to 16×16, this calculation may increase the running/computational cost and so the complexity [2]. However, as per the results discussed in Sect. 5.3, the proposed methods which benefited from the optimization algorithms and the quantization table attained a good balance between the security, image quality and secret text capacity. Perhaps, the results of the proposed methods exhibited improved solutions as compared to JQTM, non-optimal substitution method (16×16) and non-optimal substitution method (8×8). With a goal to improve the robustness and efficiency of the proposed algorithms, Approach 4 is identified and it is discussed in the following Chapter.

References

1. Almohammad, A., Ghinea, G., Hierons, R.M.: JPEG steganography: a performance evaluation of quantization tables. In: Proceedings of International Conference on Advanced Information Networking and Applications, Bradford, UK, 26–29 May 2009, pp. 471–478 (2009)
2. Bracamonte, J., Ansorge, M., Pellandini, F.: Adaptive block-size transform coding for image compression. In: IEEE International Conference on Acoustics, Speech, and Signal Processing (ICASSP-97), 21–24 April 1997, vol. 4, pp. 2721–2724 (1997)
3. Cuiling, J., Yilin, P., Lun, G., Bing, J., Xiangyu, G.: A high capacity steganographic method based on quantization table modification. Wuhan Univ. J. Nat. Sci. **16**(3), 223–227 (2011)
4. Kulkarni, A.J., Baki, M.F., Chaouch, B.A.: Application of the cohort intelligence optimization method to three selected combinatorial optimization problems. Eur. J. Oper. Res. **250**(2), 427–447 (2016)
5. Kulkarni, M.: An information hiding system using 16×16 quantization table. In: Proceedings of International Conference on Advances in Communication and Computing Technologies, Mumbai, India, 10–11 Aug 2014, pp. 1–4
6. Kulkarni, A.J., Durugkar, I.P., Kumar, M.: Cohort intelligence: a self supervised learning behavior. In: Proceedings of the 2013 IEEE International Conference on Systems, Man and Cybernetics, IEEE Computer Society, Washington, DC, USA, 13–16 Oct 2013, pp. 1396–1400
7. Langelaar, G.C., Setyawan, I., Lagendijk, R.L.: Watermarking digital image and video data: a state-of-the-art overview. IEEE Signal Process. Mag. **17**(5), 20–46 (2000)
8. Li, X., Wang, J.: A steganographic method based upon JPEG and particle swarm optimization algorithm. Inf. Sci. **177**, 3099–3109 (2007)
9. Wang, R.Z., Lin, C.F., Lin, J.C.: Image hiding by optimal LSB substitution and genetic algorithm. Pattern Recogn. **34**(3), 671–683 (2001)
10. Zhu, F.: Blocking artifacts reduction in compressed data. In: Proceedings of the 2009 International Conference on Computer Engineering and Applications, IPCSIT, 2, IACSIT Press, Singapore (2011)

Chapter 6
Improved Cohort Intelligence Optimization Algorithm (Approach 4)

Nowadays the level of information security has been enhanced by various concepts such as cryptography, steganography [9] along with nature-inspired optimization algorithms [4, 7, 8]. However, in today's world computational cost (time and function evaluations) plays a vital role in the success of any scientific method. The optimization algorithms, such as CICC and M-MRSLS were already implemented and applied for JPEG image steganography for 8×8 as well as 16×16 quantization table, respectively. Although results were satisfactory in terms of image quality and capacity, the computational time was high for most of the test images. To overcome the challenge, this research work proposed a modified version of the CI algorithm referred to as Improved CI [6]. The Improved CI was considered as a cryptography technique and implemented to generate the optimized ciphertext. The Improved CI was further employed for JPEG image steganography. Experimentation was done on the grayscale image, of size 256×256; both for 8×8 and 16×16 quantization table. The proposed work exhibited very encouraging improvements in the computational cost.

6.1 The Improved CI Optimization Algorithm

The proposed algorithm is presented in Fig. 6.1. This algorithm is an extension of CI originally presented in Kulkarni et al. [3]. It was based on the social interactions of candidates competing with one another to achieve a shared goal. The Improved CI algorithm intended to implement quality factor and luck factor as characteristics of the cohort candidate. This helped the CI algorithm to represent a more realistic scenario.

The steps of Improved CI algorithm are described below:

Step 1: Initialize the cohort with a number of candidates C. Each substitution matrix (S_c) was modeled as a cohort candidate and was viewed either as an identity matrix

© The Editor(s) (if applicable) and The Author(s), under exclusive license to Springer Nature Switzerland AG 2020

D. K. Sarmah et al., *Optimization Models in Steganography Using Metaheuristics*, Intelligent Systems Reference Library 187, https://doi.org/10.1007/978-3-030-42044-4_6

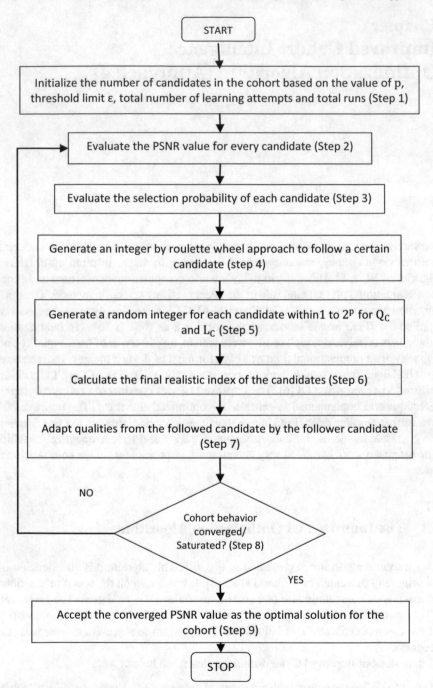

Fig. 6.1 Improved CI algorithm (approach 4)

or its variations which helped the receiver to extract the secret message by taking the inverse of the matrix.

The total number of substitution matrices for a group of p secret bits were $2^p!$. The representation of (S_c) is as follows wherein a and b are considered as rows and column index of each matrix S and the quality of each candidate is considered a value 1 lies in the matrix against each row and column index a and b respectively.

$$S_C = \{s_{ab}, 0 \le a, b \le 2^p - 1\}, c = 1, 2, \ldots, C$$

$$where\ S_{ab} = \begin{cases} 1, & if\ a\ replaces\ b \\ 0, & otherwise \end{cases} \tag{6.1a}$$

Step 2: Evaluate the PSNR value for every candidate S_c in the cohort for the same secret message.

Step 3: Evaluate the selection probability for every candidate as shown below. The value of selection probability for each candidate decided the chance of selecting itself or other candidates. The selection probability of any candidate is dependent upon the number of candidates in the cohort and was calculated by dividing the PSNR value of the individual candidate with total PSNR value calculated for all the candidates.

$$P^{Sc} = \frac{PSNR_{Sc}}{\sum_{c=0}^{C} PSNR_{Sc}} \tag{6.1b}$$

Step 4: An integer random number was generated by any candidate from within 1 to 2^p by applying the Roulette wheel approach to decide the follower and following candidate. There was an integer value selected by the candidates for a parameter p which was used to divide the secret text into p number of bits.

Step 5: Two integer random numbers were generated referred to as Q_C and L_C by every candidate from within 1 to 2^p by applying the Roulette wheel approach. The quality index Q_C indicates the possibility of adopting certain qualities of any candidate as per the generated random integer, whereas, the luck index L_C indicates the luck factor for any candidate to achieve the desired goal. In order to improve certain quality(s) of any candidate, one should have the value of three parameters viz. P^{Sc}, Q_C and L_C.

Step 6: Evaluate the final realistic index R_C of any candidate by considering the below-mentioned sub-steps:

Step 1: The multiplication of three parameters: P^{Sc}, Q_C and L_C were calculated.

Step 2: The round off operation was further applied to the output received from Sub-step 1 to achieve the nearest integer to calculate R_C.

The value of R_C signifies the number of qualities to be adapted by other candidates or themselves.

$$R_C = \| P^{Sc} \times Q_C \times L_C \| \tag{6.1c}$$

Each candidate was considered as an identity matrix and its variations. The position of each row-column of the matrix where the value 1 appears denoted the quality of the candidate. The total number of qualities of a candidate was decided by the frequency of appearing 1 in the matrix which makes its overall behavior.

Step 7: In order to adapt certain qualities from the candidates based on R_C, the number of rows of the follower candidate was replaced by the number of rows of the followed candidate(s).

Step 8: Verify whether the Cohort behavior is converged or saturated either by considering the following sub-steps:

Sub-step 1: If the maximum number of learning attempts is exhausted.
Sub-step 2: If there is no significant improvement in the behavior of every candidate in the cohort which implies no considerable difference recognized between the PSNR values evaluated for each candidate in the respective iterations. The following conditions are considered:

$$\left| \max\left(P^{S_C}\right)^n - \max\left(P^{S_C}\right)^{n-1} \right| \le \varepsilon \; and \tag{6.1d}$$

$$\left| \min\left(P^{S_C}\right)^n - \min\left(P^{S_C}\right)^{n-1} \right| \le \varepsilon \; and \tag{6.1e}$$

$$\left| \max\left(P^{S_C}\right)^n - \min\left(P^{S_C}\right)^{n-1} \right| \le \varepsilon \tag{6.1f}$$

Step 9: The final value of *PSNR* is selected as an optimal solution.

6.2   Improved CI with JPEG Image Steganography Using 8 × 8 Quantization Table

The overall embedding procedure and extraction procedure are shown in Figs. 6.1 and 6.4, respectively. For the embedding process, firstly a secret text was accepted and a JPEG image compression steganography was used to hide the secret text. The steps of JPEG image compression were explained in Sarmah and Kulkarni [8] wherein after embedding the cipher message into the cover image, RLE and DPCM were applied. EC and HC were applied further to the outcome of the previous step. To transform the secret text into an optimized cipher text Improved CI optimization algorithm was implemented and applied as a cryptographic technique. The optimized ciphertext was embedded into low-mid frequency coefficients of DCT. The frequency coefficients were selected based on the quantization table of size 8 × 8. The standard quantization table of JPEG is shown in Table 6.1. However, the quantization table as used by Li and Wang [4] was selected for this research in order to increase the secret text capacity. The modified quantization table is shown in Table 6.2.

Table 6.1 Standard JPEG quantization table

16	11	10	16	24	40	51	61
12	12	14	19	26	58	60	55
14	13	16	24	40	57	69	56
14	17	22	29	51	87	80	62
18	22	37	56	68	109	103	77
24	35	55	64	81	104	113	92
49	64	78	87	103	121	120	101
72	92	95	98	112	100	103	99

Table 6.2 JPEG quantization table (proposed by Li and Wang [4])

8	1	1	1	1	1	1	1
1	1	1	1	1	1	1	55
1	1	1	1	1	1	69	56
1	1	1	1	1	87	80	62
1	1	1	1	68	109	103	77
1	1	1	64	81	104	113	92
1	1	78	87	103	121	120	101
1	92	95	98	112	100	103	99

6.2.1 Embedding Procedure with Example

The detailed embedding procedure with illustration is explained in this section. As depicted in Fig. 6.1, all the steps for secret text transformation, optimal substitution matrix identification using Improved CI with its illustration are described under three points a, b and c respectively (Fig. 6.2).

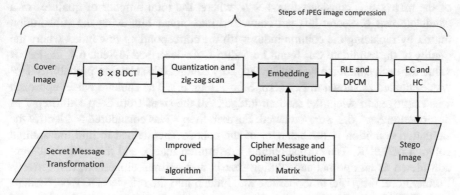

Fig. 6.2 Embedding procedure (8 × 8 quantization table)

(a) **Secret Text Transformation**:

To transform the secret text into different ciphertexts, the following steps are considered.

Step 1: Secret Text is accepted.
Step 2: Convert the secret text into binary digits.
Step 3: Form a group of p $bits$ from the output generated in Step 2.
Step 4: Convert each group of p secret bits into a decimal number. The range of decimal numbers for p secret bits is from 0 to $2^p - 1$.
Step 5: Generate different p cipher bits with respect to all the candidates to further produce different ciphertexts, correspondingly.

Since the cipher bits were to be hidden into the DCT coefficients of the JPEG image, there was a need to generate optimal ciphertext from the optimal substitution matrix. Thus, an optimization algorithm was required. The next objective was to achieve the optimal ciphertext from the optimal substitution matrix which is described in the next section.

(b) **Optimal Substitution Matrix Identification using Improved CI**

An improved CI optimization algorithm was implemented and applied to find the optimal substitution matrix. Each step of the algorithm is described in Fig. 6.1. The value of p was assumed as 2. This value was considered to illustrate the proposed work effectively. The total number of candidates' C and substitution matrices S for the value $p = 2$ are $2^p! = 2^2! = 24$. In order to demonstrate efficiently, randomly 4 candidates were selected out of 24. Total runs and threshold limit ε were considered as 40 and 0.0001, respectively. These values were considered based on the previous experimentation and analysis was done using CICC and M-MRSLS algorithm to solve the steganography problem [7, 8]. Each run was having 20 iterations. Evaluation of a stego image was done on the basis of image quality which was calculated using PSNR as mentioned in Eq. 4.4.1b.

The evaluated decimal values against p = 2 were 4 from 0 to 3; thus, the size of the matrix was considered as 4×4, where, the total number of qualities of a candidate was 4. Secret bits get converted into cipher bits using the substitution matrix by replacing the column index with the corresponding row index where the quality of the candidate was found i.e. value 1. As described in Sect. 6.4, the PSNR value for each candidate was evaluated. Any candidate's selection probability P^{Sc} was evaluated as shown in Eq. 6.1b, Step 3, Fig. 6.1. The roulette wheel approach was employed to select the random integer. All the steps from Step 1 to Step 7 as described in Sect. 6.2 were executed. Further, Step 8 was considered to identify the saturation condition in the behavior of the cohort. This helped to find the optimal solution of PSNR. The corresponding substitution matrix and the ciphertext were considered as an optimal substitution matrix and optimal ciphertext, respectively. Furthermore, the p bits of ciphertext was hidden into the selected DCT coefficients of the JPEG image. The illustration for the previous sections is explained under point (c).

(c) **Illustration**

An illustration of this algorithm is also presented below. Each step was considered with respect to a sample grayscale image Lena. A single iteration is considered here for depiction.

1. The following 4 candidates are considered for $p = 2$:

$$\begin{bmatrix} 0\,1\,0\,0 \\ 1\,0\,0\,0 \\ 0\,0\,1\,0 \\ 0\,0\,0\,1 \end{bmatrix} \quad \begin{bmatrix} 1\,0\,0\,0 \\ 0\,0\,1\,0 \\ 0\,1\,0\,0 \\ 0\,0\,0\,1 \end{bmatrix} \quad \begin{bmatrix} 0\,0\,0\,1 \\ 0\,1\,0\,0 \\ 1\,0\,0\,0 \\ 0\,0\,1\,0 \end{bmatrix} \quad \begin{bmatrix} 0\,0\,1\,0 \\ 0\,0\,0\,1 \\ 1\,0\,0\,0 \\ 0\,1\,0\,0 \end{bmatrix}$$

$$\text{(a)}\,S_1 \qquad \text{(b)}\,S_2 \qquad \text{(c)}\,S_3 \qquad \text{(d)}\,S_4$$

2. Let us assume the secret text is of 8 bits:$\{1\,1\,0\,1\,1\,0\,0\,1\}$
3. As described in Sect. 2.2.1, step 3, a group of 2 bits is formed. Thus, in total 4 groups are generated for 8 bits of secret text i.e. $\{1\,1\,0\,1\,1\,0\,0\,0\}$
4. Each pair gets converted into a decimal number:$\{3\,1\,2\,0\}$
5. The decimal values of secret bits will be substituted through different substitution matrices/candidates to generate new secret bits or cipher bits:

   (a) Substitution with respect to S_1: $\{3\,0\,2\,1\}$
   (b) Substitution with respect to S_2: $\{3\,0\,1\,2\}$
   (c) Substitution with respect to S_3: $\{2\,1\,0\,3\}$
   (d) Substitution with respect to S_4: $\{1\,3\,0\,2\}$

6. Convert each decimal value with respect to every cipher in the bits form.

   (a) Cipher bits with respect to $S_1 = \{1\,1\,0\,0\,1\,0\,0\,1\}$
   (b) Cipher bits with respect to $S_2 = \{1\,1\,0\,0\,0\,1\,1\,0\}$
   (c) Cipher bits with respect to $S_3 = \{1\,0\,0\,1\,0\,0\,1\,1\}$
   (d) Cipher bits with respect to $S_4 = \{0\,1\,1\,1\,0\,0\,1\,0\}$

7. Hide all the secret bits in the combination of p bits to the selected DCT coefficients of JPEG image as mentioned in the quantization table of Sarmah and Kulkarni [8] and evaluate their *PSNR* value with respect to all the steps (a), (b), (c) and (d) as stated in point 6.

$$\text{PSNR(a)} = 38.2109, \quad \text{PSNR(b)} = 38.2212,$$
$$\text{PSNR(c)} = 38.2356, \quad \text{PSNR(d)} = 38.2199$$

8. Selection probability with respect to the first candidate

$$P^{S_1} = \frac{\text{PSNR(a)}}{\text{PSNR(a)} + \text{PSNR(b)} + \text{PSNR(c)} + \text{PSNR(d)}}$$

9. Calculate the total probability

$$P^{S_T} = 38.2109 + 38.2128 + 38.2348 + 38.2199 = 152.8784$$

10. $P^{S_1} = \dfrac{38.2109}{152.8784} = 0.2499, \quad P^{S_2} = \dfrac{38.2212}{152.8784} = 0.2500,$

    $P^{S_3} = \dfrac{38.2356}{152.8784} = 0.2501, \quad P^{S_4} = \dfrac{38.2199}{152.8784} = 0.2500.$

11. Roulette wheel approach is applied to evaluate the follower candidate and to be followed candidate as described in below steps:

    11.1 Cumulative probability (CP) of every candidate is calculated.

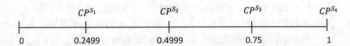

    11.2 As there are 4 candidates and every candidate is having 4 qualities; thus, every candidate will generate 4 random values through the Roulette wheel approach to follow the individual. The random values are:

$$\{0.3523, 0.8911, 0.1185, 0.6545\}$$

    11.3 As per the generated random values, the first candidate will follow the second one, the second candidate will follow the fourth one, the third candidate will follow the first one and the fourth candidate will follow the third one.

$$\text{cand}_1 \rightarrow \text{cand}_2$$
$$\text{cand}_2 \rightarrow \text{cand}_4$$
$$\text{cand}_3 \rightarrow \text{cand}_1$$
$$\text{cand}_4 \rightarrow \text{cand}_3$$

12. There is a quality index Q_i and luck index L_i associated with each candidate who decides the adaptability quality by modifying the qualities of other candidate(s) and the luck factor, respectively, to achieve the common goal. A random integer is generated for quality index and luck index between 1 and 4 for the candidates.

$$Q_1 = 2, \quad Q_2 = 3, \quad Q_3 = 4, \quad Q_4 = 1$$
$$L_1 = 3, \quad L_2 = 4, \quad L_3 = 1, \quad L_4 = 2$$

13. The final realistic index (R_i) is evaluated for every candidate as follows. The realistic index of every candidate decides the exact number of qualities to be adopted by the candidate.

$R_i = \| P^{S_i} \times Q_i \times L_i \|$ where i varies from 1to 4.

$$R_1 = \|(2 \times 3 \times 0.2499)\| = 1$$
$$R_2 = \|(3 \times 4 \times 0.2500)\| = 3$$
$$R_3 = \|(4 \times 1 \times 0.2501)\| = 1$$
$$R_4 = \|(1 \times 2 \times 0.2500)\| = 1$$

Hence,

$cand_1$ adapts a single quality of $cand_2$.

$cand_2$ adapts three qualities of $cand_4$.

$cand_3$ adapts a single quality of $cand_1$.

$cand_4$ adapts a single quality of $cand_3$.

14. In order to adapt the qualities of certain candidates, random integers are generated by every candidate in the range of 1–4 as described in the previous step 13. The generated random integers are as follows:

For $cand_1$: {3},

For $cand_2$: {1, 2, 4},

For $cand_3$: {1},

For $cand_4$: {2}

These random numbers with respect to each candidate indicate the number of rows to be replaced by the follower candidate with the followed candidate.

15. Replace the row 3 of $cand_1$ with $cand_2$, rows 1, 2, 4 of $cand_2$ with $cand_4$, row 1 of $cand_3$ with $cand_1$ and row 2 of $cand_4$ with $cand_3$ respectively.

$$
\begin{bmatrix} 0&1&0&0 \\ 1&0&0&0 \\ 0&0&1&0 \\ 0&0&0&1 \end{bmatrix}
\quad
\begin{bmatrix} 1&0&0&0 \\ 0&0&1&0 \\ 0&1&0&0 \\ 0&0&0&1 \end{bmatrix}
\quad
\begin{bmatrix} 0&0&0&1 \\ 0&1&0&0 \\ 1&0&0&0 \\ 0&0&1&0 \end{bmatrix}
\quad
\begin{bmatrix} 0&0&1&0 \\ 0&0&0&1 \\ 1&0&0&0 \\ 0&1&0&0 \end{bmatrix}
$$

(a) S_1 (b) S_2 (c) S_3 (d) S_4

(a)   row 3 of $cand_2$ or $S_2 = \{0100\}$;

$$
\text{The new candidate } S_1' = \begin{bmatrix} 0&1&0&0 \\ 1&0&0&0 \\ 0&1&0&0 \\ 0&0&0&1 \end{bmatrix}
$$

(b)   rows 1, 2, 4 of c and_4 or $S_4 = \{0010\}, \{0001\}, \{0100\}$;

$$
\text{The new candidate } S_2' = \begin{bmatrix} 0&0&1&0 \\ 0&0&0&1 \\ 0&1&0&0 \\ 0&1&0&0 \end{bmatrix}
$$

(c) row 1 of $cand_1$ or $S_1 = \{0\,1\,0\,0\}$;

$$\text{The new candidate } S_3' = \begin{bmatrix} 0 & 1 & 0 & 0 \\ 0 & 1 & 0 & 0 \\ 1 & 0 & 0 & 0 \\ 0 & 0 & 1 & 0 \end{bmatrix}$$

(d) row 2 of $cand_3$ or $S_3 = \{0\,1\,0\,0\}$;

$$\text{The new candidate } S_4' = \begin{bmatrix} 0 & 0 & 1 & 0 \\ 0 & 1 & 0 & 0 \\ 1 & 0 & 0 & 0 \\ 0 & 1 & 0 & 0 \end{bmatrix}$$

16. Repeat the same process with the new candidates unless either the cohort behavior is converged or the saturation condition is achieved with respect to 40 runs, 20 iterations.

In the next section, the embedding procedure of secret text in the JPEG grayscale image is shown with respect to the 16×16 quantization table.

6.3 Improved CI with JPEG Image Steganography Using 16×16 Quantization Table

As described by Sarmah and Kulkarni [6, 7], the capacity of the hidden secret text plays a very major role in retaining good stego image quality. Visual distortion might take place in the stego image as the number of secret bits gets increased. Thus, a solution was proposed by Almohammad et al. [1] for JPEG image compression by increasing the size of the quantization table from 8×8 to 16×16. As shown in Table 5.1, the considered number of frequency coefficients of this quantization table is more than the number presented in Table 4.1, which facilitates to increase the secret text hidden capacity. The proposed optimization algorithm Improved CI was applied to the secret text to have the optimal ciphertext. This ciphertext was hidden into the selected DCT coefficients by using a 16×16 quantization table which helped to improve the secret text capacity along with stego image quality. Figure 6.3 describes the embedding procedure with respect to the 16×16 quantization table. In this case, the cover image was divided into the blocks of size 16×16 and DCT was applied to each block to get frequency coefficients. The detailed procedure is mentioned in Sect. 6.3.1.

6.3.1 Embedding Procedure with Example

In this section, the embedding procedure as shown in Fig. 6.3 is explained in two sub-points of d and e respectively.

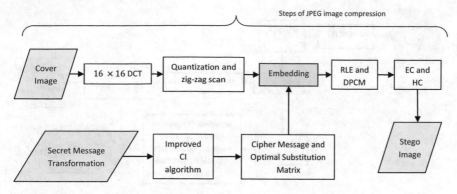

Fig. 6.3 Embedding procedure (16 × 16 quantization table) (approach 4)

(a) **Secret Text Transformation**

The description and the overall procedure of this section are the same as described in Sect. 6.2.1. As an outcome, the same number of ciphertexts was received as the number of candidates in the cohort. The next point (e) elaborates to identify the optimal substitution matrix and the optimal ciphertext by hiding the ciphertexts into some particular DCT coefficients calculated using a 16 × 16 quantization table and with the help of the proposed optimization algorithm Improved CI.

(b) **Optimal Substitution Matrix Identification using Improved CI Algorithm and illustration**

As shown in Table 5.1 [7], the positions of the selected DCT coefficients are clear from this quantization table as there are total 121 values of $1's$ lies in this table and as described above the p bits of secret text are hidden into each DCT coefficient. The entire process of finding the optimal substitution matrix and optimal ciphertext using Improved CI with respect to the 8×8 quantization table is elaborated in Sect. 6.2.1(b). Its illustration is also shown in Sect. 6.2.1(c). Moreover, the same steps are considered for this section. Once the secret text was embedded into a JPEG image at the sender end, it was important to retrieve the original secret text at the receiver end. Thus, a retrieval algorithm was required which is mentioned in Sect. 6.3.2.

6.3.2 Extraction Procedure

The objective of this algorithm was to retrieve the secret message from the stego image. The receiver accepted the stego image and the optimal substitution matrix as an input, sent by the sender. The reverse procedure of JPEG image compression and Improved CI algorithm was applied at the receiver end as shown in Fig. 6.4. In the stego image, Entropy Decoding (ED) and Huffman Decoding (HD) were applied. The reverse process of RLE and DPCE referred to as Run Length Decoding (RLD) and

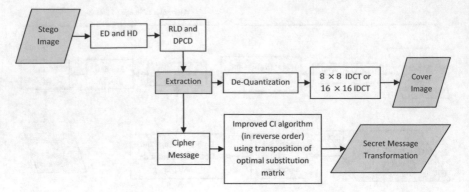

Fig. 6.4 Extraction procedure

Digital Pulse Code Demodulation (DPCD) was applied to extract the cipher message. Here, the step of extraction took place which was divided into two branches. In one branch the ciphertext was extracted and passed to the transposition of the optimal substitution matrix. To extract the secret message from the ciphertext, the Improved CI algorithm was applied in the reverse order. The other branch was used to extract the cover image. In this branch, the dequantization process was applied to the image. The output received from this step was further considered and the inverse process of DCT was applied as per the size of the quantization table to extract the cover image.

The experimental results of the proposed method are presented in Sect. 6.4. This section discusses and validates the result of the proposed work with respect to the other existing methods such as Joint Quantization Table Modification (JQTM) as described in Li and Wang [4], Non-optimal substitution method (refer Li and Wang [4] for more details), JPEG_PSO, CICC and M-MRSLS (proposed methods by Sarmah and Kulkarni [7]) optimization algorithm. Further, the results of the Improved CI algorithm was validated by considering two sizes of quantization table of 8×8 and 16×16.

6.4   Comparison and Discussion

Six grayscale images of different variety and texture were considered as the cover image for testing purposes. These images were Lena, Baboon, Woman, Boat, Girl and Gold Hill. The size of these 8-bit images was considered as 256×256 for testing. The validation of the proposed method was done with respect to different evaluation parameters such as Image quality with respect to PSNR and visual interpretation, the capacity of secret text, the standard deviation of function evaluations, and computational time. The worst, best and median values for the image quality and computational time were also evaluated. A total of 40 runs were considered. Each run was having 20 iterations. As discussed in the thesis, two quantization tables were used for size 8×8 and 16×16. Thus, each image was divided into 8×8 and 16×16 blocks respectively to have quantized DCT coefficients. The extensive analysis of the proposed method was done by comparing this with the other existing methods. These methods were JQTM (refer to Sarmah and Kulkarni [7] for more details), Non-optimal substitution, JPEG_PSO, CICC, and M-MRSLS. All the algorithms identified for comparing with the proposed work were implemented under the same circumstances. As discussed before, the implementation of CICC (variation of CI used for steganography) and M-MRSLS (variation of MRSLS) was completed by considering the 8×8 quantization table as well as 16×16 quantization table. In order to improve the results, the other variation of CI i.e. Improved CI was implemented on similar lines. The comparison for image quality with respect to different methods for all the testing images implemented using an 8×8 quantization table is shown in Table 6.3. One could observe that the result of the proposed method with respect to the JQTM and non-optimal substitution method was quite good; however, the PSNR values for the other methods were comparable. Since CICC and M-MRSLS were implemented for the 16×16 quantization table as well under the same conditions, Table 6.4 presents the comparison for image quality between the proposed method, CICC, and M-MRSLS. The excellent results were achieved for Improved CI in comparison to the other methods for all the test images. The comparative analysis for best, median and worst image quality was also completed for both the quantization

Table 6.3 Comparison: image quality (PSNR in decibel (dB))

Images	Methods (8×8 quantization table)					
	JQTM	Non-optimal substitution	JPEG_PSO	CICC	M-MRSLS	Improved CI
Lena	35.8600	37.7800	37.0600	38.2417	38.2417	38.2348
Baboon	29.5103	29.1646	31.2800	29.9376	29.9376	29.9355
Woman	39.9383	42.1203	35.7100	42.3867	42.4212	42.4019
Boat	36.2393	38.5884	36.2900	38.6128	38.5428	38.6099
Gold hill	36.63	41.0400	36.7800	41.6636	41.6362	41.6252
Girl	29.3303	29.5103	38.0200	30.1102	30.1102	30.1151

Table 6.4 Comparison: image quality (PSNR in decibel (dB))

Images	Methods (16×16 quantization table)		
	CICC	M-MRSLS	Improved CI
Lena	34.2691	34.2691	35.0870
Baboon	27.6438	27.6428	28.6272
Woman	38.3797	38.3797	39.1293
Boat	33.7763	33.7763	34.6591
Gold hill	34.2110	34.2106	34.9043
Girl	28.0231	28.0225	28.9198

Table 6.5 Comparison: best, median and worst image quality (PSNR in dB)

Images		Methods (8 × 8 quantization table)		
		CICC	M-MRSLS	Improved CI
Lena	Best	38.2417	38.2417	38.2348
	Median	38.2417	38.2275	38.2178
	Worst	38.1725	38.1725	38.2109
Baboon	Best	29.9376	29.9376	29.9355
	Median	29.9358	29.935	29.9335
	Worst	29.934	29.9335	29.9327
Woman	Best	42.4212	42.4212	42.4019
	Median	42.3867	42.4212	42.3942
	Worst	42.3867	42.3625	42.3876
Boat	Best	38.6204	38.6204	38.6099
	Median	38.6128	38.5428	39.6001
	Worst	38.5592	38.5428	38.5903
Goldhill	Best	41.6636	41.6640	41.6252
	Median	41.6636	41.6360	41.6010
	Worst	41.6225	41.6290	41.5724
Girl	Best	30.1151	30.1200	30.1151
	Median	30.1102	30.1100	30.1106
	Worst	301,097	30.1100	30.1089

tables of size 8×8 and 16×16. The presentation of this comparison is done in Tables 6.5 and 6.6 respectively.

The standard deviation of the image quality was evaluated for the proposed algorithm as well as the contemporary algorithms selected for comparison. The result analysis of this comparison with respect to two quantization tables can be depicted in Tables 6.7 and 6.8 respectively. The standard deviation of Improved CI was found very less in comparison to the other methods for all the test images which indicated good visual image quality for the proposed work. Tables 6.9 and 6.10 have shown the standard deviation of function evaluations for the methods. Very less number of function evaluations was observed for Improved CI by using both the quantization tables in comparison to CICC and M-MRSLS. The standard deviation of computational time is presented in Tables 6.11 and 6.12. As per the experimental results depicted in Tables 6.11 and 6.12, it was observed that the computational time for Improved CI was exceedingly improved than the other existing methods which made this algorithm more realistic and practical. The best, median and worst cases of computational time were also evaluated for all the test images by using both the quantization tables separately. They are presented in respective tables of Tables 6.13 and 6.14. For most of the test images, Improved CI performed exceptionally better in comparison to

Table 6.6 Comparison: best, median and worst Image quality (PSNR in dB)

Images		Methods (16×16 quantization table)		
		CICC	M-MRSLS	Improved CI
Lena	Best	34.2691	34.2691	35.0870
	Median	34.2585	34.2543	35.0830
	Worst	34.2083	34.2083	35.0747
Baboon	Best	27.6438	27.6428	28.6272
	Median	27.6398	27.6398	28.6256
	Worst	27.6371	27.6371	8.6237
Woman	Best	38.3797	38.3797	39.1293
	Median	38.3719	38.35335	39.1283
	Worst	38.3708	38.3113	39.1271
Boat	Best	33.7763	33.7763	34.6591
	Median	33.7652	33.75595	34.6565
	Worst	33.7652	33.7361	34.6562
Goldhill	Best	34.2110	34.2106	34.9043
	Median	34.1880	34.1877	34.9012
	Worst	34.1750	34.1754	34.8922
Girl	Best	28.0231	28.0225	28.9198
	Median	28.0186	28.0182	28.9148
	Worst	28.0180	28.0087	28.9128

Table 6.7 Comparison: standard deviation of PSNR

Images	Methods			
	Standard deviation of PSNR (8×8 quantization table)			
	JPEG_PSO	CICC	M-MRSLS	Improved CI
Lena	0.0300	0.0181	0.0197	0.0054
Baboon	0.0100	0.0008	0.0013	0.0006
Woman	0.0200	0.0079	0.0224	0.0039
Boat	0.0200	0.0137	0.0281	0.0039
Goldhill	0.0300	0.0138	0.0105	0.0149
Girl	0.0300	0.0012	0.0019	0.0011

CICC and M-MRSLS methods with respect to computational time without compromising image quality; however, for the other test images, the results were found comparative. The comparison with respect to secret text is hidden capacity was also analyzed for Improved CI and the other existing methods as shown in Tables 6.15 and 6.16. Each method selected for comparison considered 2 bits of secret message to be hidden in each selected DCT coefficient. As shown in Tables 6.15 and 6.16, the

Table 6.8 Comparison: standard deviation of PSNR

Images	Methods		
	Standard deviation of PSNR (16×16 quantization table)		
	CICC	M-MRSLS	Improved CI
Lena	0.0159	0.0255	0.0029
Baboon	0.0022	0.0017	0.0011
Woman	0.0034	0.0253	0.0007
Boat	0.0051	0.0144	0.0008
Goldhill	0.0108	0.0109	0.0044
Girl	0.0018	0.0055	0.0015

Table 6.9 Comparison: standard deviation of function evaluations

Images	Methods		
	Standard deviation of function evaluations (8×8 quantization table)		
	CICC	M-MRSLS	Improved CI
Lena	36	31	4
Baboon	29	32	4
Woman	33	28	5
Boat	29	32	5
Goldhill	27	31	4
Girl	35	28	4

Table 6.10 Comparison: standard deviation of function evaluations

Images	Methods		
	Standard deviation of function evaluations (16×16 quantization table)		
	CICC	M-MRSLS	Improved CI
Lena	36	31	3
Baboon	29	32	5
Woman	33	28	4
Boat	29	32	4
Goldhill	27	31	3
Girl	35	28	4

total number of 36 and 121 coefficients were selected respectively which enabled to hide a total of $2 \times 36 = 72$ bits and $2 \times 121 = 242$ bits in each block. Thus, the capacity to hide secret bits improved from 72 bits to 242 bits. The original and stego images of all the test images for Improved CI are shown in Fig. 6.5. Visual quality for cover and stego image was compared and analyzed. The simulations for Improved

Table 6.11 Comparison: standard deviation of computational time

Images	Methods					
	Standard deviation of computational time (8 × 8 quantization table)					
	Improved CI	CICC	M-MRSLS	JPEG_PSO	JQTM	Non-optimal substitution
Lena	12.9217	99.8137	79.3652	35.5000	2.4448	2.6235
Baboon	18.5803	101.2854	108.0255	35.7000	3.2807	3.2176
Woman	17.8110	86.3609	101.6035	35.5000	2.4350	2.6852
Boat	18.3571	79.5028	87.7162	35.5000	2.6226	2.8826
Goldhill	14.4584	75.0487	119.8618	35.9000	2.6253	2.8137
Girl	15.8475	99.5779	78.6989	35.7000	2.6302	3.3147

Table 6.12 Comparison: standard deviation of computational time

Images	Methods		
	Standard deviation of computational time (16 × 16 quantization table)		
	Improved CI	CICC	M-MRSLS
Lena	11.4088	71.7879	81.6882
Baboon	22.6711	144.6967	121.8759
Woman	12.2509	63.2177	92.5372
Boat	15.8575	115.7960	102.5778
Goldhill	9.71094	108.9900	77.5181
Girl	14.4777	59.8624	84.2665

CI were also captured with respect to all the test images as shown in Figs. 6.6 and 6.7. As there were 20 runs and each run was having 40 iterations, a single run was considered for all the test images for both the quantization tables on an individual basis for demonstration. The graph is plotted for the number of learning attempts as represented in the x-axis against PSNRs represented in the y-axis. Symbolic representation with respect to 4 candidates is given in each graph. The converged value for PSNR could be easily identified for all the candidates in every graph of Figs. 6.6 and 6.7.

6.5 Summary

A novel algorithm is proposed by combining JPEG steganography and Improved CI optimization algorithm which was applied to six grayscale test images of size 256 × 256. Improved CI is another variation of CI in which each candidate in the cohort envisions practically and tries to achieve their goal based on two properties: (1) The capability of the candidate to adapt certain quality(s) in others, and (2)

Table 6.13 Comparison: computational time for the best, median and worst case

Images		Methods (8×8 quantization table)		
		Improved CI	CICC	M-MRSLS
Lena	Best	89.3991	207.3000	41.2890
	Median	57.8943	265.9000	259.7620
	Worst	43.6493	520.500	354.388
Baboon	Best	122.5231	312.6016	90.6541
	Median	77.47805	633.4719	233.4110
	Worst	57.2253	641.8447	467.0896
Woman	Best	44.4479	185.9100	71.3700
	Median	59.8593	238.2200	286.7000
	Worst	121.2800	506.3700	485.2000
Boat	Best	118.1789	248.9599	58.9551
	Median	63.2759	453.5200	368.7303
	Worst	47.2845	459.7481	368.7303
Goldhill	Best	92.5388	234.4130	58.3548
	Median	62.5160	379.5108	236.3060
	Worst	46.8513	474.4410	500.6340
Girl	Best	112.1929	188.2520	89.3751
	Median	64.4952	323.2340	89.3751
	Worst	47.9593	483.4530	379.3276

Luck factor of the candidate. By adding these two properties CI was modified to Improved CI which makes this algorithm more realistic. The experimental result validated this algorithm with respect to the other existing algorithms. Improved CI was implemented by using 8×8 as well as a 16×16 quantization table. It was clear from the results that the computational time for this method was significantly less and was reduced by a minimum of 79% and a maximum 82% in comparison to CICC and minimum 80% and a maximum 83% in comparison to M-MRSLS by using 8×8 quantization table. Further, for (16×16) quantization table, the computational time for Improved CI was reduced by a minimum of 76% and a maximum of 91% as compared to CICC and minimum 81% and a maximum of 87% as compared to M-MRSLS. Image quality for certain test images was also improved and the capacity of the secret text was increased by using (16×16) quantization table. However, the doors are still open to working on steganalysis techniques to verify the security of the proposed work. This work can further be extended by controlling access rights for unauthorized users while distributing digital content by issuing one secret key [2]. Furthermore, annotation data or metadata could be easily added in a media database system which could further help to implement a keyword-based movie scene retrieving system efficiently [5]. The vulnerability analysis for all the stego images is completed in the next chapter.

Table 6.14 Comparison: computational time for the best, median and worst case

Images		Methods (16 × 16 quantization table)		
		Improved CI	CICC	M-MRSLS
Lena	Best	89.3991	207.3000	41.2890
	Median	57.8943	265.9000	259.7620
	Worst	43.6493	520.5000	354.3880
Baboon	Best	122.5231	312.6016	90.6541
	Median	77.4781	633.4719	233.4110
	Worst	57.2253	641.8447	467.0896
Woman	Best	44.4479	185.9100	71.3700
	Median	59.8593	238.2200	286.7000
	Worst	121.2800	506.3700	485.2000
Boat	Best	43.6775	222.3972	50.5418
	Median	60.1056	373.1210	204.1573
	Worst	102.6362	527.4827	439.7497
Goldhill	Best	42.5182	273.6350	49.8291
	Median	57.1709	479.6340	170.9780
	Worst	84.4008	525.1040	309.8740
Girl	Best	43.1138	272.3617	221.9240
	Median	60.4108	489.6918	298.2660
	Worst	100.5941	564.7522	512.2000

Table 6.15 Comparison: capacity (bits)

Capacity (bits)	Methods				
	(8 × 8 quantization table)				
	JQTM	JPEG_PSO	CICC	M-MRSLS	Improved CI
Selected DCT coefficients for hiding	26	36	36	36	36
Number of bits to be hidden per DCT coefficient	2	2	2	2	2
Hiding capacity per block	52	72	72	72	72

Table 6.16 Comparison: capacity (bits)

Capacity (bits)	Methods		
	(16 × 16 quantization table)		
	Improved CI	CICC	M-MRSLS
Selected DCT coefficients for hiding	121	121	121
Number of bits to be hidden per DCT coefficient	2	2	2
Hiding capacity per block	242	242	242

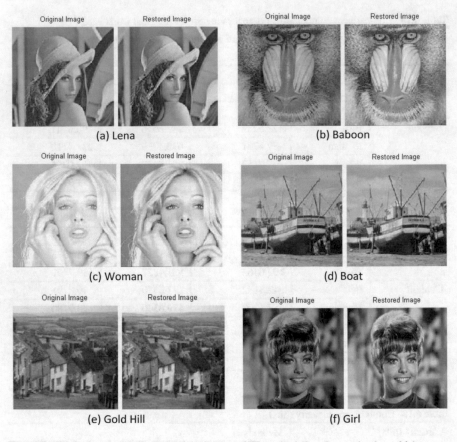

Fig. 6.5 Original and stego-images of the improved CI method (8 × 8 quantization table)

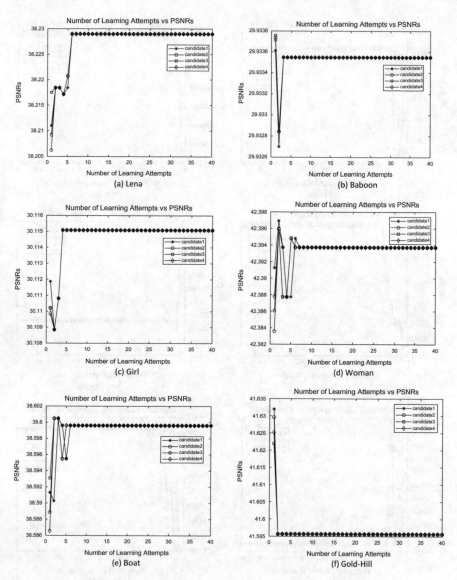

Fig. 6.6 Improved CI simulations-20 runs (8×8 quantization table)

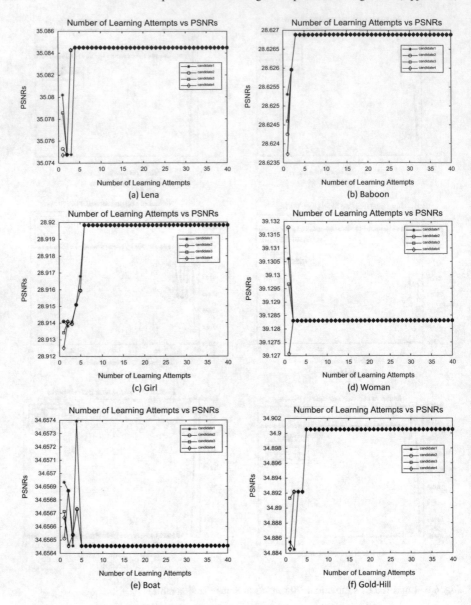

Fig. 6.7   Improved CI simulations-20 runs (16 × 16 quantization table)

References

1. Almohammad, A., Ghinea, G., Hierons, R.M.: JPEG steganography: a performance evaluation
    of quantization tables. In: Proceedings of International Conference on Advanced Information
    Networking and Applications, Bradford, UK, 26–29 May 2009, pp. 471–478 (2009)

2. Kawaguchi, E., Maeta, M., Noda, H., Nozaki, K.: A model of digital contents access control system using steganographic information hiding scheme. In: Proceedings of Information Modeling and Knowledge Bases XVIII, Trojanovice, Czech Republic, 29 May–2 June 2006, pp. 50–61 (2006)

3. Kulkarni, A.J., Durugkar, I.P., Kumar, M.: Cohort intelligence: a self supervised learning behavior. In: Proceedings of the 2013 IEEE International Conference on Systems, Man and Cybernetics, IEEE Computer Society, Washington, DC, USA, 13–16 October 2013, pp. 1396–1400 (2013)

4. Li, X., Wang, J.: A steganographic method based upon JPEG and particle swarm optimization algorithm. Inf. Sci. **177**, 3099–3109 (2007)

5. Patel, N.S., Abowd, D.G.: The contextcam: automated point of capture video annotation. In: Davies, N., Mynatt, E.D., Siio, I. (eds.) UbiComp 2004: Ubiquitous Computing, UbiComp 2004, Lecture Notes in Computer Science (LNCS), vol. 3205, pp 301–318. Springer, Berlin (2004)

6. Sarmah, D.K., Kulkarni, A.J.: Improved cohort intelligence—a high capacity, swift and secure approach on JPEG image steganography. J. Inf. Secur. Appl. **45**, 90–106 (2019)

7. Sarmah, D., Kulkarni, A.J.: JPEG based steganography methods using cohort intelligence with cognitive computing and modified multi random start local search optimization algorithms. Inf. Sci. **430–431**, 378–396 (2018)

8. Sarmah, D., Kulkarni, A.J.: Image steganography capacity improvement using cohort intelligence and modified multi random start local search methods. Arab. J. Sci. Eng. **43**(8), 3927–3950 (2018)

9. Sarmah, D., Bajpai, N.: Proposed system for data hiding using cryptography and steganography. Int. J. Comput. Appl. **8**(9), 7–10 (2010)

Chapter 7
Steganalysis on All Approaches/Vulnerability Analysis of Stego Image(s)

Stego image quality, secret text embedding capacity, computational time and security are the main challenges involved for steganography methods. The proposed methods referred to as Cohort Intelligence with Cognitive Computing (CICC) [22], Modified-Multi Random Start Local Search (M-MRSLS) [23] and Improved Cohort Intelligence (Improved CI) [21] exhibited encouraging results against stego image quality, secret text embedding capacity and computational time as described in the previous chapters 4, 5 and 6. These methods were also tested against different steganalysis attacks [4, 7] such as visual detection, structural detection, and statistical detection to validate the security aspect of these algorithms. Image steganalysis is explained in the following section which has been divided into three subsections. Each subsection describes different approaches—visual detection [10], structural detection [5] and statistical detection [5] of image steganalysis. Comparison and result analysis are presented in Sect. 7.2. The summary of the entire section is explained in Sect. 7.3.

7.1 The Objective of Image Steganalysis

Through steganalysis, the attacker can identify the presence of hidden text in digital media. The main goal of a steganalyst is to identify whether a secret text is present or not in the stego media [4]. Many researchers tried to categorized steganalysis attacks [11, 26] based on steganalysis approaches such as statistical steganalysis and feature-based steganalysis. In the proposed work, image steganalysis was used to identify the secret text from an image. There are different ways through which image analysis [7, 8, 26] can be done for which distinct methods and tools are available to detect Steganography. Image steganalysis is divided into two broad categories: (a) Target steganalysis [4, 7], (b) Blind steganalysis [1, 15, 18]. The target steganalysis is used whenever a particular steganography embedding algorithm and a particular type

D. K. Sarmah et al., *Optimization Models in Steganography Using Metaheuristics*, Intelligent Systems Reference Library 187, https://doi.org/10.1007/978-3-030-42044-4_7

of image is focused. Different statistical properties are analyzed under this type of steganalysis which aims to provide accurate results. For example: Out Guess attack [16], MB1 attack [15], LSB Matching Steganography attack [26], YASS attack [19], JSTEG [1], F5 [25], etc. On the other hand, Blind steganalysis can be applied for any type of image and any steganography embedding algorithm. Though the result of this technique is not very accurate, it is most commonly used by the researchers due to its simplicity. For example, the blind image steganalysis can be employed for binary similarity [2], wavelet-based analysis [27], feature-based analysis [6], etc. For the proposed work, Target steganalysis was used which is further divided into three well-known detection techniques, viz: Visual detection [3, 4], Structural detection [5, 12] and Statistical detection [5, 15]. These techniques are discussed in the section below.

7.1.1   Visual Detection

The visual detection could be analyzed by seeing the difference between the cover image and the stego image. The possible difference could be either small distortions in stego image or some visible cyclic patterns in the stego image. It might create suspicion for steganalyst to further analyze the stego image to disclose any secret text. This is referred to as a known-cover attack [9]. In this work, visual detection was done for all the proposed methods and inspected for all the test images.

7.1.2   Structural Detection

By modifying the structure of a stego image, one can identify the difference between the cover image and stego image, which in turn provides a ground to steganalyst to identify the secret text. In this technique, the file properties of the cover image and stego image were observed and analyzed. All the test images implemented on CI, CICC and Improved CI for (16×16) quantization tables were selected to analyze this attack. The size of the image was selected as an attribute for this attack and the percentage difference in size between the cover image and stego image was observed.

7.1.3   Statistical Detection

The statistical detection technique was performed by analyzing the image pixels. It is categorized into two types: Spatial domain analysis and Transforms domain analysis. Since JPEG image compression was used in the proposed work, transform domain analysis was performed for statistical detection. Histogram analysis [17] was performed on the DCT coefficients. It is based on the difference between the histogram values of the DCT coefficients. Histogram finds a relationship between intensity values and its frequency. A graph for the image histogram was drawn and the difference in the normalized histogram for cover and stego image was calculated. The secret text can be suspected through histogram by noticing brightness, contrast and pixel distribution of the cover image and stego image.

The next section discusses the results and presents its analysis for techniques such as visual detection, structural detection, and statistical detection.

7.2 Comparison and Discussion

The three techniques as mentioned in the Sect 7.1.1, 7.1.2 and 7.1.3 are considered for the steganalysis exploration. The proposed methods referred to as CICC, M-MRSLS and Improved CI were used with (16 × 16) quantization table. The results presented in Fig. 5.9a and 6.5a was considered to analyze the results for the visual detection steganalysis technique. By reviewing these images, one could easily identify the distortions between the cover image and the stego image. The six test images were used for the experimentation. Human eyes could perceive a difference of brightness between the cover image and stego image for the images Baboon and Girl. The less brightness was observed for the Woman stego image in comparison to its cover image for CICC and M-MRSLS methods. However, it could be difficult to identify the distortions by human eyes for the rest of the test images as these methods were not supported by visual steganalysis. Furthermore, the structural detection steganalysis was applied to all the test images in which the 'image size' attribute was selected for analysis. The methods considered for this approach of steganalysis were CI, CICC and Improved CI with (16 × 16) quantization table. The results of this analysis are presented in Tables 7.1, 7.2 and 7.3 for the methods CI, CICC, and Improved CI, respectively. As shown in Table 7.1 for CI method, the percentage of image size difference for all the test images was found less than 5%; however, for CICC and Improved CI, as shown in the respective tables, the percentage of image size difference for the test images Lena and Woman exceeded more than 5%. This proved stego image obtained by CI is more secure than using CICC and Improved CI for structural steganalysis.

Further, Statistical steganalysis was applied to all the test images for CI, CICC and Improved CI. Their results are presented in Tables 7.4, 7.5 and 7.6, respectively. Image Histogram tool was used for this technique and the percentage difference between the cover image and stego image was calculated in the normalized histogram. An insignificant percentage difference was observed for most of the images for all the three methods. The maximum difference was detected for the Woman image i.e. 0.0083 (approximate) and the minimum difference was perceived for the Gold hill Image i.e. 0.0011 (approximate) for each method, which indicated that the proposed methods are secured against this attack.

Table 7.1 Structural steganalysis (CI)

Test Images	Property-Cover Image	Property-Stego Image	Percentage Size Difference
Baboon			2.80%
Boat			2.80%
Girl			4.68%
Goldhill			1.92%

(continued)

Table 7.1 (continued)

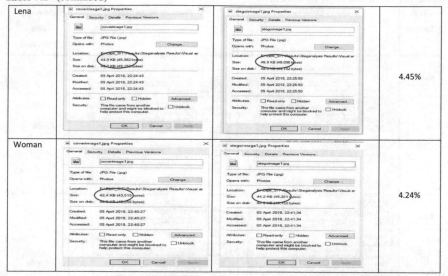

7.3 The Characteristic Difference Between the Well-Known Optimization Algorithms with the Proposed Algorithms

As discussed in Sect. 7.3, there are many recognized metaheuristic algorithms. Genetic Algorithm (GA) and Particle Swarm Optimization (PSO) have gained a lot of importance for several researchers to solve various types of problems. In this section, the comparative difference of characteristics of the strong metaheuristic algorithms such as GA and PSO with the proposed algorithms such as CI, CICC, and Improved CI is explained. The advantages and disadvantages are also described for each algorithm.

GA [14] is based on Darwinian Theory which works on the concept of survival of the fittest among individuals from a population. This is an evolutionary algorithm based on genetic operators such as selection, crossover, mutation, etc. Exploration and exploitation in GA are carried forward by crossover and mutation respectively which helps the solution to jump from local optima to global optima. Also, GA is used to solve discrete problems and applied for combinatorial problems. However, it has been observed that GA is computationally exhaustive as there is a large size of the population required to reach the solution in the close neighborhood of optima solution. On the other hand, Swarm optimization (SO) algorithms [24] such as PSO [24], ACO [20], etc. imitates the social behavior of living creatures. In PSO, the internal velocity of the particle decides its position. PSO is a continuous technique and cannot be applied to solve combinatorial problems. Although it does not have two operators such as crossover and mutation, PSO generates new solutions in the close vicinity of optima solution. There are two populations used in PSO: (i) P_{best}

Table 7.2 Structural steganalysis (CICC)

Test Images	Property-Cover Image	Property-Stego Image	Percentage Size Difference
Baboon			3%
Boat			1.93%
Girl			2.4%
Goldhill			1.28%

(continued)

Table 7.2   (continued)

Lena			7.7%
Woman			6.5%

i.e. the best position of the particle and (ii) particle's current positions which allow exploration and exploitation. In PSO, a particle is considered as an agent where each agent observes the status of other agents. The best agent/particle allows the other particles to update their status. This enables the solution to converge in a faster way. This algorithm is based on a one-way sharing mechanism. In GA, there are no sensing capabilities to sense their neighborhood. The number of genes can be selected based on a fitness function. Whereas PSO is easy to implement as there is a limited number of parameters. Further, there is no decoding or encoding involved. Thus, computational efficiency is also found high for this algorithm. However, in traditional PSO, diversity can be improved by combining this with other metaheuristics algorithms to enhance its local searchability.

As described in Sect. 7.3, the CI algorithm [13] is inspired by the social behavior of candidates. A self-supervised learning behavior can be observed in the cohort by interactions and competitions amongst the candidates to achieve the desired goal. CI is having the capability of handling constraints in various problems. However, this is still a big challenge for nature/bio-inspired techniques which support to solve the unconstrained problem. The performance of such techniques reduces drastically when included in the problem with constraints. In CI, the learning ability of every candidate from others is based on a certain probability. This helps the candidate to adopt a single quality from any other candidate for a single iteration. It could be repeated iteratively to improve cohort behavior. Worse behavior may also be followed by the

Table 7.3 Structural steganalysis (improved CI)

Test Images	Property-Cover Image	Property-Stego Image	Percentage Size Difference
Baboon			1.07%
Boat			3.01%
Girl			5.67%
Goldhill			2.77%

(continued)

Table 7.3   (continued)

candidates who allow the solution to jump from possible local optima as similar to the SA. Due to this uncertainty, the possibility of exploration gets increased. There are few important parameters involved in this algorithm when applied to steganography problem such as a number of candidates and sampling interval reduction factor which plays an important role to decide the quality of solutions. However, tuning of these parameters is required mechanically to get a better solution. This algorithm is used, applied to several types of problems such as discrete, continuous, constrained, unconstrained, etc. and validated with the contemporary algorithms. However, it has been observed during experimentation to solve the image steganography problem that there is a high computational cost involved when (i) any of the suggested parameters get change, (ii) image size increases and (iii) secret text hiding capacity increases. Also, there was a slow convergence exhibited by CI. Further, there is another aspect of consideration for improving solutions by increasing the capability of exploitation between the candidates to help every candidate to make an effective step.

CICC was proposed to work on the limitations of CI. In CICC, a cognitive approach was induced amongst the candidates who enhance the exploitation between the candidate solutions. Every candidate observes the other candidate in the cohort iteratively and if the candidate's solution in the current iteration is found better than its behavior of the previous iteration, then due to the CC approach, the candidate accepts the solution of current iteration else the previous solution is maintained. The convergence behavior and quality of solutions of this algorithm were found better than CI. However, there was no significant improvement in computational efficiency. As similar to CICC, in Improved CI, exploitation capabilities are enhanced by adding two more parameters such as quality index and luck index along with the number of existing parameters. In CI and CICC, the selection probability of every candidate decides the follower candidate for them. However, Improved CI allows the candidate to follow certain candidates based on three parameters: (i) selection probability, (ii) quality index, and (iii) luck index which improves the quality of the solution. This algorithm is found so effective to solve image steganography problems for reducing the computational cost, generating effective solutions very close to its optima solution and security.

Table 7.4 Statistical steganalysis (CI)

Test Images	Histogram-Cover Image	Histogram-Stego Image	Percentage Difference in Normalized Histogram (MSE)
Baboon	Number of Gray Levels Vs Pixel Intensities	Number of Gray Levels Vs Pixel Intensities	0.0040051
Boat	Number of Gray Levels Vs Pixel Intensities	Number of Gray Levels Vs Pixel Intensities	0.0021445
Girl	Number of Gray Levels Vs Pixel Intensities	Number of Gray Levels Vs Pixel Intensities	0.0036444

(continued)

Table 7.4 (continued)

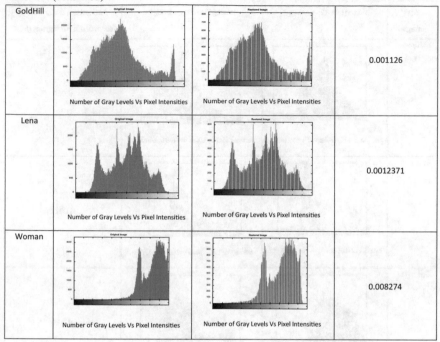

GoldHill	Number of Gray Levels Vs Pixel Intensities	Number of Gray Levels Vs Pixel Intensities	0.001126
Lena	Number of Gray Levels Vs Pixel Intensities	Number of Gray Levels Vs Pixel Intensities	0.0012371
Woman	Number of Gray Levels Vs Pixel Intensities	Number of Gray Levels Vs Pixel Intensities	0.008274

7.4 Summary

Different steganalysis techniques such as visual detection steganalysis, structural steganalysis, and statistical steganalysis were applied to validate the security of the proposed methods i.e. CI, CICC, and Improved CI with (16×16) quantization table. The result analysis as shown in the above section revealed that the proposed methods were secured against the mentioned steganalysis attacks for most of the test images. However, these methods could be tested and verified against some advanced/latest steganalysis technique(s) used for JPEG image steganography.

Table 7.5 Statistical Steganalysis (CICC)

Test Images	Histogram-Cover Image	Histogram-Stego Image	Percentage Difference in Normalized Histogram (MSE)
Baboon	Number of Gray Levels Vs Pixel Intensities	Number of Gray Levels Vs Pixel Intensities	0.004
Boat	Number of Gray Levels Vs Pixel Intensities	Number of Gray Levels Vs Pixel Intensities	0.0021
Girl	Number of Gray Levels Vs Pixel Intensities	Number of Gray Levels Vs Pixel Intensities	0.036
Goldhill	Number of Gray Levels Vs Pixel Intensities	Number of Gray Levels Vs Pixel Intensities	0.0011
Lena	Number of Gray Levels Vs Pixel Intensities	Number of Gray Levels Vs Pixel Intensities	0.0012

(continued)

Table 7.5 (continued)

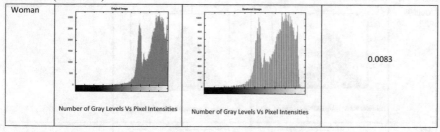

Woman	Original Image — Number of Gray Levels Vs Pixel Intensities	Restored Image — Number of Gray Levels Vs Pixel Intensities	0.0083

Table 7.6 Statistical steganalysis (improved CI)

Test Images	Histogram-Cover Image	Histogram-Stego Image	Percentage Difference in Normalized Histogram (MSE)
Baboon	Original Image — Number of Gray Levels Vs Pixel Intensities	Restored Image — Number of Gray Levels Vs Pixel Intensities	0.004
Boat	Original Image — Number of Gray Levels Vs Pixel Intensities	Restored Image — Number of Gray Levels Vs Pixel Intensities	0.0021
Girl	Original Image — Number of Gray Levels Vs Pixel Intensities	Restored Image — Number of Gray Levels Vs Pixel Intensities	0.0036

(continued)

Table 7.6 (continued)

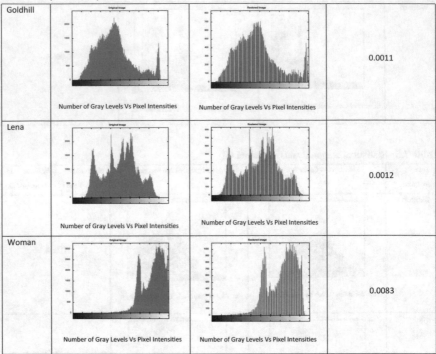

Goldhill	Number of Gray Levels Vs Pixel Intensities	Number of Gray Levels Vs Pixel Intensities	0.0011
Lena	Number of Gray Levels Vs Pixel Intensities	Number of Gray Levels Vs Pixel Intensities	0.0012
Woman	Number of Gray Levels Vs Pixel Intensities	Number of Gray Levels Vs Pixel Intensities	0.0083

References

1. Attaby, A.A., Ahmed, M.F.M.M., Alsammak, A.K.: Data hiding inside JPEG images with high resistance to steganalysis using a novel technique: DCT-M3. Ain Shams Eng. J. **9**(4), 1965–1974 (2018)
2. Avcıbas, I., Kharrazi, M., Memon, N., Sankur, B.: Image steganalysis with binary similarity measures. EURASIP J. Appl. Sig. Process. **17**, 2749–2757 (2015)
3. Banerjee, S., Ghosh, B.R., Roy, P.: JPEG steganography and steganalysis—a review. In: Satapathy, S., Biswal, B., Udgata, S., Mandal, J. (eds.) Proceedings of the 3rd International Conference on Frontiers of Intelligent Computing: Theory and Applications (FICTA) 2014, Advances in Intelligent Systems and Computing, vol. 328, pp. 175–187. Springer, Cham (2015)
4. Chandramouli, R., Kharrazi, M., Memon, N.: Image steganography and steganalysis: concepts and practice. In: Kalker, T., Cox, I., Ro, Y.M. (eds.) Digital Watermarking, IWDW 2003, Lecture Notes in Computer Science (LNCS), vol. 2939, pp. 35–49. Springer, Berlin (2004)
5. Douglas, M., Bailey, K., Leeney, M., Curran, K.: An overview of steganography techniques applied to the protection of biometric data. Multimedia. Tools Appl. **77**(13), 17333–17373 (2018)
6. Fridrich, J.: Feature-based steganalysis for JPEG images and its implications for future design of steganographic schemes. In: International Workshop on Information Hiding, Information Hiding, Lecture Notes in Computer Science (LNCS), vol. 3200, pp. 67–81 (2004)
7. Fridrich, J.: Methods for tamper detection in digital images. In: ACM Workshop on Multimedia and Security, Orlando, FL, 30–31 Oct, 1999, pp. 19–23 (1999)

8. Hedieh, S.: Image steganalysis using Artificial Bee Colony algorithm. J. Exp. Theor. Artif. Intell. **29**(5), 949–966 (2017)
9. Heys, H.M., Tavares, S.E.: Known plaintext cryptanalysis of tree-structured block ciphers. Electron. Lett. **31**(10), 784–785 (1995)
10. Hussain, M., Wahab, A.W.A., Idris, Y.I.B., Ho, A.T.S., Jung, K.: Image steganography in spatial domain: a survey. Sig. Process. Image Commun. **65**, 46–66 (2018)
11. Johnson, N.F., Jajodia, S.: Steganalysis of images created using current steganography software. In: Aucsmith, D. (eds.) Information Hiding, IH 1998, Lecture Notes in Computer Science (LNCS), vol. 1525, pp 273–289. Springer, Berlin (1998)
12. Ker, A.D.: A general framework for structural steganalysis of LSB replacement. In: Information Hiding, Lecture Notes in Computer Science, vol. 3727, pp. 296–311 (2005)
13. Kulkarni, A.J., Durugkar, I.P., Kumar, M.: Cohort intelligence: a self supervised learning behavior. In: Proceedings of the 2013 IEEE International Conference on Systems, Man and Cybernetics, 13–16 Oct 2013, pp. 1396–1400. IEEE Computer Society, Washington, DC, USA (2013)
14. McCall, J.: Genetic algorithms for modelling and optimisation. J. Comput. Appl. Math. **184**(1), 205–222 (2005)
15. Nissar, A., Mir, A.H.: Classification of steganalysis techniques: a study. Digit. Signal Proc. **20**(6), 1758–1770 (2010)
16. Oplatkova, Z., Holoska, J., Zelinka, I., Senkerik, R.: Detection of steganography inserted by outguess and steghide by means of neural networks. In: AMS '09 Proceedings of the 2009 Third Asia International Conference on Modelling & Simulation, May 25–29, 2009, pp. 7–12. IEEE Computer Society Washington, DC, USA (2009)
17. Qin, K., Xu, K., Liu, F., Li, D.: Image segmentation based on histogram analysis utilizing the cloud model. Comput. Math Appl. **62**(7), 2824–2833 (2011)
18. Rabee, A.M., Mohamed, M.H., Mahdy, Y.B.: Blind JPEG steganalysis based on DCT coefficients differences. Multimed. Tools Appl. **77**(6), 7763–7777 (2018)
19. Sadasivam, S., Moulin, P.: On estimation accuracy of desynchronization attack channel parameters. IEEE Trans. Inf. Forensics Secur. **4**(3), 284–292 (2009)
20. Santis, R.D., Montanari, R., Vignali, G., Bottani, E.: An adapted ant colony optimization algorithm for the minimization of the travel distance of pickers in manual warehouses. Eur. J. Oper. Res. **267**(1), 120–137 (2018)
21. Sarmah, D.K., Kulkarni, A.J.: Improved cohort intelligence-a high capacity, swift and secure approach on JPEG image steganography. J. Inf. Secur. Appl. **45**, 90–106 (2019)
22. Sarmah, D., Kulkarni, A.J.: JPEG based steganography methods using cohort intelligence with cognitive computing and modified multi random start local search optimization algorithms. Inf. Sci. **430–431**, 378–396 (2018)
23. Sarmah, D., Kulkarni, A.J.: Image steganography capacity improvement using cohort intelligence and modified multi random start local search methods. Arab. J. Sci. Eng. **43**(8), 3927–3950 (2018)
24. Wang, Z., Zhang, J., Yang, S.: An improved particle swarm optimization algorithm for dynamic job shop scheduling problems with random job arrivals. Swarm Evol. Comput. **51**, 100594 (2019)
25. Westfeld, A.: F5-A steganographic algorithm: high capacity despite better steganalysis. In: Proceedings of the 4th Information Hiding Workshop, LNCS, vol. 2137, pp. 289–302 (2001)
26. Zhihua, X., Xingming, S., Wei, L., Jiaohua, Q., Feng, L.: JPEG Image steganalysis using joint discrete cosine transform domain features. J. Electron. Imaging **19**(2), 023006 (2010)
27. Zong, H., Liu, F., Luo, X.: Blind image steganalysis based on wavelet coefficient correlation. Digit. Investig. **9**(1), 58–68 (2012)

Chapter 8
Conclusions and Future Recommendations

Nowadays computer networks play a vital role and demand secured data transfer. Cryptography [16] and Steganography [16] are the two important sciences work in this area. There is always a possibility of attack by cryptanalyst [1] and/or steganalyst [4] or any unintended receivers on the sensitive information, which creates a need to enhance the growth of algorithms in the domain of information security. Information hiding as a branch of information security has various applications in military communications [11], anti-criminal [5], digital forensics [12], business [8], etc. This chapter discusses the summary of the proposed work and the recommendation(s) for the future. The authors focused on developing image steganography techniques using socio inspired optimization algorithm. Three optimization algorithms: Cohort Intelligence with Cognitive Computing (CICC) [14], Modified-Multi Random Start Local Search (M-MRSLS) [15] and Improved Cohort Intelligence (Improved CI) [13] were proposed, developed and applied to JPEG image steganography [10] for 8×8 and 16×16 quantization table. The results were validated for the important evaluation parameters like stego image quality, secret text embedding capacity, secret text security and computational time.

8.1 Conclusions

For the first time, the methods CICC and M-MRSLS were applied in the Image processing domain and implemented by using six grayscale test images of size 256 $\times$ 256. The implementation of the proposed methods was completed to address four major concerns for image steganography i.e. capacity in bits, quality of the stego image, computational time and security. The proposed methods derived an optimal substitution matrix by CICC and M-MRSLS to transform the secret messages and then hid the secret message into the cover-image through a modified JPEG

D. K. Sarmah et al., *Optimization Models in Steganography Using Metaheuristics*,
Intelligent Systems Reference Library 187,
https://doi.org/10.1007/978-3-030-42044-4_8

quantization table. The comparative results are discussed (Chap. 4) underscored that the solution to our methods with other cotemporary methods is comparable or even some times better than the other methods. Furthermore, the focus was to improve the embedding secret text capacity by retaining good image quality. CICC and M-MRSLS were applied to JPEG image steganography using a modified 16 × 16 quantization table. The proposed methods achieved larger message capacity and better image quality than JQTM, JPEG_PSO [10] and non-optimal substitution method (Chap. 5). Moreover, these methods provided high security to extract the secret message. Also, the PSNR values of the proposed methods were observed and analyzed, which validated its improved image quality over the other methods for most of the test images. Though the proposed methods: CICC, M-MRSLS and Improved CI had many advantages, few limitations were also being observed. The computational time for these methods was in the little higher side while comparing with other parallel methods. By observing this as an area of improvement, a compatible version of CI relevant to image steganography referred to as Improved CI was proposed and developed. This was another variation of CI in which each candidate in the cohort envisions practically and tries to achieve their goal based on two properties: 1. The capability of the candidate to adapt certain quality(s) in others, and 2. Luck factor of the candidate. By adding these two properties, CI was modified to Improved CI as mentioned in Chap. 6 which made this algorithm more practical. Improved CI was implemented by using 8 × 8 as well as a 16 × 16 quantization table. The evaluated computational time for this method was found significantly less. It was reduced by a minimum of 79% and a maximum of 82% in comparison to CICC and a minimum of 80% and a maximum of 83% in comparison to M-MRSLS by using an 8 × 8 quantization table. Moreover, for *the* 16 × 16 quantization table, the computational time for Improved CI was reduced by a minimum of 76% and a maximum of 91% as compared to CICC and minimum 81% and a maximum *of* 87% as compared to M-MRSLS. Image quality for certain test images was also improved and the capacity of the secret text was increased by using a 16 × 16 quantization table. Furthermore, the robustness of the proposed methods was observed so that these methods could not reveal any secret information during some famous steganalysis attack. Few well-known steganalysis attacks i.e. visual detection, structural detection, and statistical detection were applied to the methods: CI (16 × 16), CICC (16 × 16) and Improved CI (16 × 16). The result analysis was done in Chap. 7. The respective simulations for statistical steganalysis were also carried out to validate the security aspect of these proposed methods.

8.2   Recommendations for Future Work with Examples

Though the proposed methods achieved stability between image quality, secret text embedding capacity, security and computational time, there was a scope of improvement. For future work, the authors would make an effort to apply more steganalysis attacks to further validate the security of the secret text. The enrichment could also be done by applying the optimization algorithms in neural-based steganography [6], machine learning-based steganography [2], deep steganography [3], etc. Moreover, this work could be explored using DWT instead of DCT which could further be enhanced by selecting the random pixel dynamically in each block for hiding. In addition, this work could also be extended by controlling access rights for unauthorized users while distributing digital content by issuing one secret key [7]. Also, the proposed optimization algorithms: CICC, M-MRSLS and Improved CI, could be explored solving other applications to highlight its effectiveness. Although the proposed optimization algorithms were used to find the optimal substitution matrix along with the optimal ciphertext, their behavior was identified as a cryptography algorithm. There could be a combination of any strong cryptography algorithm such as AES [16], blowfish [9], etc. with the proposed optimization algorithm to make the system more secure and complex to withstand any attack. The secret text embedding capacity could also be improved by increasing more number of bits for hiding in each block without distorting the stego image quality.

References

1. Ahmad, M., Khan, I.R., Alam, S.: Cryptanalysis of image encryption algorithm based on fractional-order lorenz-like chaotic system. In: Satapathy S.C. et al. (eds.), Emerging ICT for Bridging the Future, Advances in Intelligent Systems and Computing, Proceedings of the 49th Annual Convention of the Computer Society of India CSI, vol. 2, pp. 381–388 (2015)
2. Atee, H.A., Ahmad, R., Noor, N.M., Rahma, A.M.S., Ajeroudi, Y.: Extreme learning machine based optimal embedding location finder for image steganography. PLoS ONE **12**(2), 1–23 (2017)
3. Baluja, S.: Hiding images in plain sight: deep steganography. In: Proceedings of the 31st Conference on Neural Information Processing Systems, 4–9 Dec 2017, pp. 1–11. Long Beach, CA, USA (2017)
4. Bohme, R.: Principles of Modern Steganography and Steganalysis. In: Advanced Statistical Steganalysis, Information Security and Cryptography, vol. 0, pp. 11–77 (2010)
5. Choraś, M., Kozik, R., Flizikowski, A., Hołubowicz, W., Renk, R.: Cyber threats impacting critical infrastructures. In: Managing the Complexity of Critical Infrastructures, Studies in Systems, Decision, and Control (SSDC), vol. 90, pp. 139–161 (2017)
6. El-Emam, N.N.: New Data-hiding Algorithm based on Adaptive Neural Networks with Modified Particle Swarm Optimization. Comput. Secur. **55**, 21–45 (2015)
7. Kawaguchi, E., Maeta, M., Noda, H., Nozaki, K.: A model of digital contents access control system using steganographic information hiding scheme. In: Proceedings of Information Modeling and Knowledge Bases XVIII, Trojanovice, Czech Republic, 29 May–2 June 2006, pp. 50–61 (2006)
8. Kong, H., Kim, T., Kim, J.: An Analysis on Effects of Information Security Investments: a BSC Perspective. J. Intell. Manuf. **23**(4), 941–953 (2012)
9. Krishnamurthy, G.N., Ramaswamy, V., Leela, G.H.: Performance enhancement of blowfish algorithm by modifying its function. In: Sobh, T., Elleithy, K., Mahmood, A., Karim, M. (eds.) Innovative Algorithms and Techniques in Automation, Industrial Electronics and Telecommunications, pp. 241–244. Springer, Dordrecht (2007)
10. Li, X., Wang, J.: A steganographic method based upon JPEG and particle swarm optimization algorithm. Inf. Sci. **177**, 3099–3109 (2007)

11. Musheng, Y., Yu, Z.: The research of intelligent monitoring system based on digital image processing. In: Second International Conference on Intelligent Computation Technology and Automation, 10–11 Oct 2009. IEEE, Hunan, China (2009)
12. Peterson, G.: Forensic analysis of digital image tampering. In: IFIP International Conference on Digital Forensics, Advances in Digital Forensics, IFIP—The International Federation for Information Processing (IFIPAICT), vol. 194, pp. 259–270 (2005)
13. Sarmah, D.K., Kulkarni, A.J.: Improved cohort intelligence—a high capacity, swift and secure approach on JPEG image steganography. J. Inf. Secur. Appl. **45**, 90–106 (2019)
14. Sarmah, D., Kulkarni, A.J.: JPEG based steganography methods using cohort intelligence with cognitive computing and modified multi random start local search optimization algorithms. Inf. Sci. **430–431**, 378–396 (2018)
15. Sarmah, D., Kulkarni, A.J.: Image steganography capacity improvement using cohort intelligence and modified multi random start local search methods. Arab. J. Sci. Eng. **43**(8), 3927–3950 (2018)
16. Sarmah, D., Bajpai, N.: Proposed system for data hiding using cryptography and steganography. Int. J. Comput. Appl. **8**(9), 7–10 (2010)

Printed in the United States
By Bookmasters